PARTNERSHIPS IN MARINE RESEARCH

Science of Sustainable Systems

PARTNERSHIPS IN MARINE RESEARCH

Case Studies, Lessons Learned, and Policy Implications

Volume Editors

GUILLERMO AUAD
Bureau of Safety and Environmental Enforcement, US Department of the Interior, Sterling, VA, United States

FRANCIS K. WIESE
Stantec Consulting Services, Inc., Anchorage, AK, United States

Series Editors

BRIAN FATH

DAN FISCUS

ELSEVIER

Elsevier
Radarweg 29, PO Box 211, 1000 AE Amsterdam, Netherlands
The Boulevard, Langford Lane, Kidlington, Oxford OX5 1GB, United Kingdom
50 Hampshire Street, 5th Floor, Cambridge, MA 02139, United States

Copyright © 2022 Elsevier Inc. All rights reserved.

No part of this publication may be reproduced or transmitted in any form or by any means, electronic or mechanical, including photocopying, recording, or any information storage and retrieval system, without permission in writing from the publisher. Details on how to seek permission, further information about the Publisher's permissions policies and our arrangements with organizations such as the Copyright Clearance Center and the Copyright Licensing Agency, can be found at our website: www.elsevier.com/permissions.

This book and the individual contributions contained in it are protected under copyright by the Publisher (other than as may be noted herein).

Notices
Knowledge and best practice in this field are constantly changing. As new research and experience broaden our understanding, changes in research methods, professional practices, or medical treatment may become necessary.

Practitioners and researchers must always rely on their own experience and knowledge in evaluating and using any information, methods, compounds, or experiments described herein. In using such information or methods they should be mindful of their own safety and the safety of others, including parties for whom they have a professional responsibility.

To the fullest extent of the law, neither the Publisher nor the authors, contributors, or editors, assume any liability for any injury and/or damage to persons or property as a matter of products liability, negligence or otherwise, or from any use or operation of any methods, products, instructions, or ideas contained in the material herein.

Library of Congress Cataloging-in-Publication Data
A catalog record for this book is available from the Library of Congress

British Library Cataloguing-in-Publication Data
A catalogue record for this book is available from the British Library

ISBN: 978-0-323-90427-8

For information on all Elsevier publications visit our website at https://www.elsevier.com/books-and-journals

Publisher: Candice Janco
Acquisitions Editor: Louisa Munro
Editorial Project Manager: Chris Hockaday
Production Project Manager: Kumar Anbazhagan
Cover Designer: Matthew Limbert

Typeset by TNQ Technologies

Cover Credit: The book cover design is by Franco Auad who used an underwater photograph taken on July 4, 2019 by Vincent Legrand off Terceira Island, Azores, Portugal. It shows short-beaked common dolphin (Delphinus delphis) feeding on bait ball. One of the dolphins is bubbling to stress the school of fish. The image of the Argo Deep float was kindly provided by the Instrument Development Group at Scripps Institution of Oceanography. The image emphasizes the importance of collaborations as dolphins work together to feed on the fish, while the fish are schooling and working together to increase their chances of survival. Meanwhile, scientists on the airplane and a research vessel work together to gather observations and learn about the ocean

Contents

Contributors	xi
About the editors	xv
Acknowledgments	xvii
Introduction	xix

1. The Bering Sea Project — 1
Francis K. Wiese

Introduction	1
Background	1
Building the partnership	2
Program implementation	4
Program management	7
Program findings	9
Sustainability	10
Lessons learned	11
Conclusions	13
Acknowledgment	15
References	15

2. Belmont Forum partnerships — 17
Erica L. Key

Introduction	17
Partnership processes	18
Implementing the partnerships	22
Outcomes and lessons learned	27
Acknowledgment	31

3. The Nansen Legacy: pioneering research beyond the present ice edge of the Arctic Ocean — 33
Paul Wassmann

Introduction	33
Background	34
Norway's Arctic future?	35
First steps toward the Nansen Legacy project and goals	37

v

Challenges and activities 38
Presenting the plan to industry and international partners 41
Quality control of the research plan 41
Funding the Nansen Legacy project 42
Recruitment for a sustainable Arctic future 43
The significance of cooperation, division of labor, and focus for solving grand national and international research questions 43
Lessons learned? 46
How will the future Arctic look like and how should we adapt to research efforts? 47
Acknowledgments 49
References 49

4. The Argo Program   53
Dean Roemmich, W. Stanley Wilson, W. John Gould, W. Brechner Owens, Pierre-Yves Le Traon, Howard J. Freeland, Brian A. King, Susan Wijffels, Philip J.H. Sutton and Nathalie Zilberman

Introduction 53
Argo's roots in WOCE 55
The formation of Argo's multi-national partnership 57
Implementation of the global Argo array 59
International governance: Argo and JCOMM 61
The Argo data management system 63
The sustainability of Argo 64
Conclusion 66
Acknowledgments 68
References 68

5. The Marine Arctic Ecosystem Study partnership: planning, implementation and lessons learned   71
Guillermo Auad and Francis K. Wiese

Introduction 71
Project planning 71
Project implementation 73
Lessons learned 77
Conclusions 83
Acknowledgments 85
References 85

6. **Partnering with the public: The Coastal Observation and Seabird Survey Team**   87
Julia K. Parrish, Hillary Burgess, Jaqueline Lindsey, Lauren Divine, Robert Kaler, Scott Pearson and Jane Dolliver

Introduction	87
Marine citizen science	88
COASST as a case study	90
Motivation to start the program	90
Organizational evolution	92
Lessons learned	101
Conclusion - partnerships in citizen science	104
Acknowledgments	104
References	104

7. **Long-term sustainability of ecological monitoring: perspectives from the Multi-Agency Rocky Intertidal Network**   109
Lisa Gilbane, Richard F. Ambrose, Jennifer L. Burnaford, Mary Elaine Helix, C. Melissa Miner, Steven Murray, Kathleen M. Sullivan and Stephen G. Whitaker

Introduction	109
Foundations of MARINe	112
Sustainable practices	117
Strengths of the partnership	125
Acknowledgments	126
References	127

8. **Deep-water study partnerships: characterizing and understanding the ecological role of deep corals and chemosynthetic communities in the Gulf of Mexico and northwest Atlantic**   131
Gregory S. Boland

Introduction	131
Partnering agency introductions	135
Chemo III	137
Lophelia II	142
Atlantic Canyons	147
Summary comments	151
Acknowledgments	152
References	153

9. Adaptation to repetitive flooding: expanding inventories of possibility through the co-production of knowledge   155
Elizabeth K. Marino, Annie Weyiouanna and Julie Raymond-Yakoubian

Introduction/background	155
Project setup and goals	158
Project implementation	160
Ongoing lessons	163
Conclusion: a more collaborative science	164
Acknowledgments	164
References	164

10. Lessons learned from nine partnerships in marine research   167
Francis K. Wiese, Guillermo Auad, Elizabeth K. Marino and Melbourne G. Briscoe

Introduction	167
Analysis	168
Resilience and sustainability of partnerships	176
Conclusions	179
Acknowledgments	180
References	180

11. Research partnerships and policies: a dynamic and evolving nexus   183
James J. Kendall, Jr., Elizabeth K. Marino, Melbourne G. Briscoe, Rodney E. Cluck, Craig N. McLean and Francis K. Wiese

Introduction	183
Research policies and partnerships	184
Case studies	187
Policy sources and impacts	189
The future of research policies	191
Recommendations for research policies	195
Acknowledgments	196
References	196

12. Global marine biodiversity partnership   199
Francis K. Wiese and Guillermo Auad

Preface	199
Introduction	199
Fertile Ground	200

The Partnership	203
Operational Phase	209
Epilogue	211
Acknowledgments	213
References	213

Index *217*

Contributors

Richard F. Ambrose
University of California, Los Angeles, CA, United States

Guillermo Auad
Office of Policy and Analysis, Bureau of Safety and Environmental Enforcement, U.S. Department of the Interior, Sterling, VA, United States

Gregory S. Boland
Bureau of Ocean Energy Management (Retired), Sterling, VA, United States

Melbourne G. Briscoe
OceanGeeks LLC, Alexandria, VA, United States

Hillary Burgess
School of Aquatic and Fishery Sciences, University of Washington, Seattle, WA, United States

Jennifer L. Burnaford
California State University, Fullerton, CA, United States

Rodney E. Cluck
U.S. Department of the Interior, Bureau of Ocean Energy Management, Sterling, VA, United States

Lauren Divine
Aleut Community of St. Paul Island Ecosystem Conservation Office, St. Paul, AK, United States

Jane Dolliver
College of the Environment, University of Washington, Seattle, WA, United States

Howard J. Freeland
Institute of Ocean Sciences, Fisheries and Oceans Canada, Sidney, BC, Canada

Lisa Gilbane
Bureau of Ocean Energy Management, Camarillo, CA, United States

W. John Gould
National Oceanography Centre Southampton, Southampton, United Kingdom

Mary Elaine Helix
Bureau of Ocean Energy Management (Retired), Camarillo, CA, United States

Robert Kaler
Migratory Birds, Alaska Region, U.S. Fish and Wildlife Service, Anchorage, AK, United States

James J. Kendall, Jr.
U.S. Department of the Interior, Bureau of Ocean Energy Management, Anchorage, AK, United States

Erica L. Key
Belmont Forum Secretariat, Montevideo, Uruguay

Brian A. King
National Oceanography Centre Southampton, Southampton, United Kingdom

Pierre-Yves Le Traon
Mercator-Ocean International, Toulouse, France and Ifremer, Plouzane, France

Jaqueline Lindsey
School of Aquatic and Fishery Sciences, University of Washington, Seattle, WA, United States

Elizabeth K. Marino
Oregon State University - Cascades, Bend, OR, United States

Craig N. McLean
National Oceanographic and Atmospheric Administration, U.S. Department of Commerce, Silver Spring, MD, United States

C. Melissa Miner
University of California, Santa Cruz, CA, United States

Steven Murray
California State University, Fullerton (Emeritus), CA, United States

W. Brechner Owens
Woods Hole Oceanographic Institution, Falmouth, MA, United States

Julia K. Parrish
School of Aquatic and Fishery Sciences, University of Washington, Seattle, WA, United States

Scott Pearson
Washington Department of Fish and Wildlife, Olympia, WA, United States

Julie Raymond-Yakoubian
Kawerak, Inc., Nome, AK, United States

Dean Roemmich
Scripps Institution of Oceanography UCSD, San Diego, CA, United States

Kathleen M. Sullivan
California State University, Los Angeles, CA, United States

Philip J.H. Sutton
National Institute of Water and Atmospheric Research, Wellington, New Zealand

Paul Wassmann
Department of Arctic and Marine Biology, Faculty for Biosciences, Fisheries, and Economics, UiT - The Arctic University of Norway, Tromsø, Norway

Annie Weyiouanna
Bering Strait School District, Shishmaref, AK, United States

Stephen G. Whitaker
National Park Service, Channel Islands, Ventura, CA, United States

Francis K. Wiese
Stantec Consulting Services, Inc., Anchorage, AK, United States

Susan Wijffels
Woods Hole Oceanographic Institution, Falmouth, MA, United States

W. Stanley Wilson
National Oceanic and Atmospheric Administration, Washington D.C., United States

Nathalie Zilberman
Scripps Institution of Oceanography UCSD, San Diego, CA, United States

About the editors

Guillermo Auad is the Senior Research Coordinator and a Science Advisor at the US Department of the Interior's Bureau of Safety and Environmental Enforcement. Earlier, he was the chief of the physical and chemical sciences branch at the Department's Bureau of Ocean Energy Management. Before 2010 and for over a decade, he was a faculty member at Scripps Institution of Oceanography, University of California, San Diego, and an Adjunct Professor of Oceanography at Palomar College. He has created national and international partnerships including an award-winning project between the United States and Canada on Arctic marine ecosystems as well as multi-nation partnerships addressing ocean sustainability, through the Belmont Forum. He was one of the US Government lead reviewers of the IPCC Fifth Assessment Report and a contributing author to the Third National Climate Assessment. Since 2010, he has focused on project management of different interdisciplinary studies used to inform decisions on offshore energy. Guillermo has co-authored national policies on Arctic research and several publications ranging from multidisciplinary studies using observations and models, to the application of socio-ecological resilience concepts for effective resource management and adaptation. He received his PhD in Oceanography from Scripps Institution of Oceanography.

Francis Wiese is a Senior Principal within Stantec's Environmental Services Group and serves as Stantec's Technical Leader for Marine research. Francis brings 28 years of experience working in the coastal and marine environment throughout the world, designing, implementing, and managing large interdisciplinary, multiinstitutional science programs in the Arctic, North Atlantic, North Pacific, Bering Sea, Gulf of Alaska, and Gulf of Mexico, with additional experience in the North Sea, Caribbean, Antarctic, and the Galapagos. Francis has worked for and with academia, government, nonprofits, and industry, is a technical reviewer for over 20 international journals, and serves on a variety of national and international science panels, committees, and working groups. He has extensively focused on environmental impacts as a result of anthropogenic stressors, climate change, socioecological resiliency, system science, marine spatial planning, coastal erosion, marine shipping, environmental policy, and adaptive management. He recently contributed a book chapter focused on a vision for a sustainable Arctic, integrating key global steps across society that are needed to reach such a future. Francis believes that collaboration and communication is key, and as such is a prolific public speaker who enjoys thinking outside the box to solve complex real-world issues. Francis earned his PhD in Conservation Biology at Memorial University of Newfoundland, Canada.

Acknowledgments

We are grateful to Jeff Green (Stantec) who, at the MARES international meeting of October of 2019 (Seattle), initially encouraged us to write about that partnership. Subsequently, his idea evolved into this book. We are also greatly thankful for the vision and enthusiastic dedication of every single author of every chapter and their remarkable contributions where they shared their direct experiences in building complex and successful partnerships. Their contributions are extraordinary examples of how social systems such as partnerships can drive sustainable efforts to useful and actionable new understanding and knowledge.

GA is also thankful to Molly Madden of the Bureau of Safety and Environmental Enforcement, for her support during the preparation of this book. FKW is grateful to his wonderful wife Sarah, who over the years supported and inspired the many ideas and put up with the long weekend hours that were spent on completing this book.

Introduction

Melbourne G. Briscoe[a], Francis K. Wiese[b], Guillermo Auad[c]

[a]OceanGeeks LLC, Alexandria, VA, United States; [b]Stantec Consulting Services, Inc., Anchorage, AK, United States; [c]Office of Policy and Analysis, Bureau of Safety and Environmental Enforcement, U.S. Department of the Interior, Sterling, VA, United States

"Partnerships" is a widely used and misused word. It is used to mean anything from two people talking about the work they are doing; to the relationship between a funding agency and a researcher (National Research Council, 1999), to several universities and industries working together on a joint project with jointly agreed goals and overall leadership, perhaps even with the money supporting the effort flowing through just one of the organizations to all the others. Historically, "partnerships" has meant two-way working collaborations, or alliances, or cooperating in some way. Some (Carnwell and Carson, 2009) have even used *partnerships* to define the roles of the organizations ("who we are"), and *collaborations* to describe the roles of the people ("what we do").

In this book, we use "partnerships" to mean at least two people, groups, organizations, or sectors working together so there is at least a two-way exchange of knowledge that benefits both partners. We focus, however, on more intense and comprehensive partnerships, especially those that include several kinds of partners and a moderate-to-high level of interdependence; these, historically, are the high-output and most exciting kinds of partnerships, and the likely future of partnerships.

The literature on partnerships attempts to parse the various kinds of partnerships into categories that are distinguished by the intensity or closeness of the partners (Table 1). These categories should be viewed more like a sliding scale rather than a rigorous delineation of the different levels of partnerships, but it clearly shows the spectrum and many of the distinguishing attributes of partnerships. Carnwell and Carson (2009), writing in the context of health and social care, argue for "three-way partnerships" and "in which there is joint agreement about what services should be provided and by whom, with joint employment, community development and teamwork seen as means of breaking down existing professional barriers and responding to local needs."

Table 1 Continuum of collaboration.

	Continuum of collaboration definitions			
	Networking	Cooperation	Coordination	Collaboration
Relationship	Not deliberate	Only mutual agreement	More formal agreement	Deliberately designed
Mission/ Goals	No common goals	Work together on joint goals; no commonly defined mission structure or planning effort	Work together on program-specific goals; more compatible missions	Solve common problems; solutions emerge from dealing constructively with difference; mutual benefit
Risk	Low risk	Limited risk	Limited risk	High risk
Resource sharing	Exchange of information	Some resources and rewards shared	Some resources and rewards shared	Shared risks, responsibilities, and rewards
Investment	Short term	Limited	Limited	Sustained relationship and effort; more durable and pervasive
Process	None	Focused	Focused	Emergent

After McKendall, V.J., 1996. Factors Facilitating Interorganizational Collaboration. Doctoral dissertation, University of Minnesota.

Motivations to partner

There are three compelling and one sociological reason to partner (Briscoe, 2008):

1. The problem being addressed is too big or too broad for you (or your organization) to tackle alone.
2. You need to share resources and/or assets.
3. Sponsor mandates.
4. It can be more fun to work with other people.

Sociologists (e.g., Merton cf Calhoun, 2010) have argued that science, or the acquisition of new knowledge more broadly, is a creative tension between competition and cooperation. The competition arises from "the institutional accent on originality" (i.e., what has the person/organization done?) versus the "collaboration of past and present generations," which is his phrase for "Standing on the Shoulders of Giants." In addition, researchers continuously cooperate in other ways, such as making judgments on the validity of a paper, on promotions, and recognitions, thus creating additional dynamics that could influence partnership building.

At the time Merton was developing his views of science in the 1960s, partnerships were few and far between, except among two or three colleagues in the same institution. There, the typical motivation to partner was often a combination of #1 and #4 above; like minds enjoyed working together, and they could share the pieces of a problem without overly competing.

Observational marine science—with its needs for ships, instruments, technicians—and the nature of marine science itself—which is often addressed within a single discipline but more properly needs the collaboration of multiple disciplines, very strongly has driven the field toward motivations #1 and #2 above. Funding sources, not wanting to have to deal with dozens of proposals to cover the dozens of partners on a major project, have gravitated toward #3 above, and so here we are today, with large cooperative/collaborative projects, instigated by group-defined objectives and sponsor-mandated collaborations. Some history of this evolution follows.

History of partnerships in marine research

Modern scientific oceanography and oceanographic expeditions unquestionably began with the Challenger Expedition[1] over 1872—1876. The British Admiralty supplied the ship, which the Royal Society refit at a cost of (today's US dollars) $10M. Wyville Thompson, from the University of Edinburgh, was the chief scientist (today's term, it was not used then); he directed a team of six scientists and staff aboard including naturalist John Murray (also Scottish, but born in Canada) and an official artist, J.J. Wild. Although there were many earlier oceanographic expeditions—notably the voyage of the Beagle with Charles Darwin in 1831—1836—the Challenger Expedition was the first to plan for a wide variety of observations covering multiple science disciplines. Observations included depth, samples of the bottom, bottom water, bottom temperature, and bottom fauna, mid-water

---

[1] https://www.challenger-society.org.uk/History_of_the_Challenger_Expedition.

temperature, fauna, and water samples, meteorology, and currents. This was done at 362 stations covering almost 69,000 miles (111,000 km). Under the guidance of John Murray, the collected samples were examined by over 100 scientists from multiple disciplines. The results were published in 50 volumes between 1885 and 1895.

By our definition, however, the Challenger Expedition was barely a partnership—lacking many of the characteristics of Table 1—but it was a meaningful collaborative endeavor between Thompson and Murray, with the Royal Society acting as the funder and the British Admiralty providing services in-kind.

Ledford (2015) suggests that team science began in earnest in the 1970s, primarily due to libraries beginning to stockpile journal subscriptions and improving researcher's access to journals in alternative fields. Prior to this time, science was very much divided into clear disciplines. Steele (1983) argues that institutions encouraged this as an easy management scheme and exacerbated the problem of addressing societal needs ("users" in our terminology) by separating researchers from applications personnel.

Knauss (2000) has a different view; he posits that the International Decade of Ocean Exploration (IDOE) beginning in 1970[2] was the source of multi-investigator, culturally sophisticated programs in oceanography and that the IDOE programs contributed to the "sociology of oceanography." He states: "By 1970, many of those with extensive seagoing experience were ready to embrace the concept of a multifaceted, multi-investigator approach to oceanographic field work. We believed it was the most cost-effective way to attack problems that were often not as well defined as we liked to suggest in our proposals. Multi-investigator programs followed naturally, and several were in the planning stage at the start of the IDOE: the Geochemical Ocean Sections Study (GEOSECS), the Mid-Ocean Dynamic Study (MODE), and the Climate: Long-Range Investigation Mapping and Predication (CLIMAP) program."

Jennings (2000) examined most of the large oceanographic programs sponsored by the United States National Science Foundation (NSF) during 1950–1980, including those of the IDOE years. He categorizes them as "big science," meaning multiscience, multiinvestigator, and multiinstitutional projects. The large efforts he calls out were the International Geophysical Year (IGY) during 1956–1959; the International Indian

---

[2] IDOE was developed at the US National Science Foundation (NSF) consequent to the United Nations's 1998 declaration of the International Year of the Ocean.

Ocean Expedition (IIOE) during 1962—1967, and the IDOE during 1971—1980. Jennings argues that the structure and management at the NSF were critical to the formation and maintenance of these large efforts.

We are currently entering the United Nations Decade of Ocean Science for Sustainable Development (2021—2030)[3], which might engender another surge in "big science," in this case under the theme: *The Science We Need for the Ocean We Want*.

There is general agreement with Jennings about "big science." Smith (1989) says, "Big science is characterized by large multidisciplinary teams, a division of labor, team commitments, agreements and negotiations on common purposes, and hierarchical organization." Even though Smith is talking about the development of the Space Telescope at NASA, this sounds familiar.

Not called out specifically in the literature but emerging as wisdom now that the 1970s are half a century behind us, is the parallel theme of the rapidly increasing availability of information at that time and the ease of sharing it. The first computerized online search system was DIALOG, from Lockheed Corporation, which started in 1966 and became widely available in 1972. You no longer had to have a large library locally available to you; all you needed was a computer terminal and a password. The 1970s were a surge of interest in multidisciplinary, multiinstitutional, multinational work. Communications were difficult, consisting mostly of telephone, telex and fax machines. The 1980s brought email (Kubany, 2008) and the 1990s brought the Internet and browsers and home computers. In retrospect, we never could have seen it coming: the confluence of three major currents—marine science turning its attention to global problems spanning the disciplines and the oceans; observational tools that could survive years in the ocean coupled with multiple satellites above viewing the subject synoptically; and communications that allowed knitting it all together, thus removing distance between investigators and turning the entire globe into a *Collaboratory*. Partnerships were not much wanted in the first three-quarters of the 20th century, but they were unavoidable in the final quarter.

By our definitions, the "big science" programs at NSF were certainly partnerships, usually fitting reasons #1-2-3 above, if not necessarily reason 4. There were reported issues of who was in charge, who was to get credit, partners changing their minds on what they wanted to focus on, and the inevitable money issues. Following McKendall's parsing of partnerships,

---

[3] https://www.oceandecade.org/.

most were cooperation or coordination, and not intense collaborations characterized by "if you fail, I fail."

Mazur and Bokyo (1981) applied social-science interview methodology to focus on the "big science" programs of the IDOE years, specifically on five projects: CLIMAP[4], MODE-1[5], GEOSECS[6], CEPEX[7], and MANGANESE NODULES[8]. These projects were chosen to represent both successes (the first three) and not so much (the remaining two), based on evaluations by the scientists involved in the projects. There were some factors common to both the successful and the unsuccessful projects, but some factors were distinct, namely:

1. "The successful projects were initiated by the scientists themselves, prompted by scientific questions. The less successful projects were stimulated more by perceived societal problems regarding pollution and natural resources and by funding opportunities to study these problems. The successful projects started with scientifically interesting questions, but the scientists on less successful projects were given a basic topic and had to think of interesting things to do."

2. "Other factors which differentiated more and less successful projects were related in some way to the operation of a core group of leaders. First, the very existence of such a core, tied closely to the more prominent PIs and persistently active throughout most of the life of the project (although individual members would change), was a characteristic of the successful projects but not of the less successful ones."

3. "Another characteristic of the successful core groups, particularly of their scientific directors, is that they were usually able to work out a good division of labor, with an administrator who freed them of many nonscientific project chores."

These are not earth-shaking conclusions, but they do lend credence to the idea that the leadership of a partnership project is critical, and that team engagement in the science or topic of the project is equally important.

---

[4] https://en.wikipedia.org/wiki/Climate:_Long_range_Investigation,_Mapping,_and_Prediction.
[5] https://science.sciencemag.org/content/185/4147/246.
[6] https://www.sciencedirect.com/science/article/abs/pii/0167508784906926.
[7] https://doi.org/10.1017/S0376892900004628.
[8] https://www.annualreviews.org/doi/abs/10.1146/annurev.ea.04.050176.001305.

Life phases in partnering

Partnerships are social complex systems that have distinct phases akin to those initially identified by Holling (1986) that have been applied to the dynamics of ecosystems, as well as social, economic, and political processes (Fath et al., 2015). We distinguish among four phases in the life of a partnership,
- Growth (creation, formation, and getting started)
- Conservation (keeping the project going, operating routinely during the project)
- Collapse/Release (marks the end of a cycle and the commencement of another one. In a partnership this could be a major change, planned or not, that triggers new group dynamics)
- Reorganization (new connectors are built and new and old elements are linked).

In the context of partnerships, we define
- ***The Growth phase*** as the formation phase of its initial structure, the seeking of partners (connectivity), the finding of funds (resources), the development of shared goals, and the initiation of the project.
- ***The Conservation phase*** as the status quo stage of the project during which it reaches high productivity levels; it is during this phase when the project becomes fully operational and when desired standards of productivity are met. On occasion, in this stage buffers or reserves can be created in anticipation of more challenging times.
- ***The Collapse or Release phase*** marks the end of the maximum productivity stage and completed deliverables are sent out to funders, the public and/or partners. This phase or stage is fast and could also include an internal crisis within the project which is often characterized by loss of resources or a decrease in connectivity; therefore, resilience is needed to overcome these challenges. Sometimes a fast assessment period characterizes the last part of the Release phase that then transitions into the Reorganization phase.
- ***The Reorganization phase*** is often characterized by a re-hardwiring of the system connecting new and old elements such as partners and institutions and their associated resources. This stage is one where improvisation, and/or emerging new leaders, and/or creativity and intuition can become very valuable. This transition is usually fast as the (social) system navigates in pursuit of its goals and stability. In its

intermediate and later stages, reorganization often incorporates new information or knowledge, queried from the inside of the partnership or brought onboard by new participants. If the reorganization and revitalization of the project are not possible or not undertaken, then collapse is the default result.

Although collapse may be the "right" result in many instances (it may well be time to move on to other challenges), it seems that continued long observational programs in the marine sciences and continued attention on the evolution and changes in a marine system are often the norm — if only because the partnership participants would prefer just to continue doing what they have been doing. This suggests that attention to the Collapse/Reorganization phase is needed perhaps already during the initial phases of the project itself. Two early questions the project might ask itself include: "What are the possible results of our project that might suggest the need for continuation—in some form—after the designated end of the project?" And, if continuation is needed, "Do we need to do anything differently during the project to make the continuance of the project more probable?" These questions elevate the importance for project leaders and managers to assess and prepare not only for the next stage but also for all others. Along these lines, Fath et al. (2015) note in their navigational recipes for complex social systems, that managers and leaders must prepare and revise their strategies in anticipation of the other three adaptive phases, regardless of which one they are currently situated.

Case studies of marine research partnerships

Across the globe, scientists conduct fragmented investigations often addressing problems where their components are interconnected with elements outside their primary scope. Budget limitations often set a limit to the scope of different research projects. Yet "big science" is needed to help tackle the big challenges of our time (e.g., climate change, loss of habitat and biodiversity, food security, energy transition, social and environmental justice). Even while standing on the shoulders of previous efforts and taking advantage of modern communication, observation, and analytical tools, building and implementing partnerships to tackle large scopes is difficult and time—consuming.

The purpose of this book is to examine nine different marine research partnerships to provide lessons learned and suggest recipes to benefit those aiming to construct similar partnerships in the future or those looking for practices to help keep their partnerships sustainable. These partnership examples are contained in Chapters 1–9.

In addition to being accounts of fascinating marine research projects in their own right that have substantially added to our understanding of ocean ecosystems and processes, we use these case studies as data from which to extract key elements contributing to the success and sustainability of the partnership (Chapter 10), and from which we informally hypothesize that the evolution and life of a given partnership can be characterized by the four stages introduced above. In the process, we highlight the importance of the social processes of socioecological system resilience underlying scientific research and partnering. Likewise, in Chapter 11, we define and examine research policies that may have aided or hindered these nine partnership efforts and ask, "what would an idealized research-policy landscape look like?" if the goal is to help foster "big science" and address grand challenges. Finally, Chapter 12 explores a hypothetical future partnership in which the lessons learned herein are applied and the needed research-policies explored in the previous chapters that have become a reality and have led to a highly successful partnership that helped restore global marine biodiversity.

The nine case studies that follow are presented in no particular order and span from nearshore to offshore habitats, the surface to deep-sea environments, from tropical to polar oceans, from regional to global scales, all motivated by climate change, environmental policy, resource management, ecosystem monitoring needs, and community concerns. Some of them were initiated by the researchers, some by the funding agencies, and some were hybrids of both approaches. Their success stories used different partnering models and were chosen to illustrate effective Coordination or Collaboration efforts, so their experiences may support future marine research partnership endeavors.

Acknowledgment

We wish to acknowledge the many partnerships that have occurred during our careers, from which experiences we have learned and have tried to coalesce that knowledge into this Introduction and this book. The people who have contributed to this journey are many, and we have tried over the years to thank them all personally. In retrospect, some of the important people and lessons only became apparent with the passage of time. We regret not knowing this sooner. The views and opinions expressed in this chapter are those of the authors and do not necessarily reflect the official policy or position of their employers or any other agency or organization.

References

Briscoe, M.G., 2008. Collaboration in the ocean sciences: best practices and common pitfalls. Oceanography 21 (3), 58–65.

Calhoun, C., 2010. Robert K. Merton: Sociology of Science and Sociology as Science. Columbia University Press, New York, p. 336.

Carnwell, R., Carson, A., 2009. The concepts of partnership and collaboration. p3-19. In: Carnwell, R., Buchanan (Eds.), Effective Practice in Health, Social Care and Criminal Justice, Second Edition. Open University Press, McGraw-Hill, Berkshire, England, p. 360.

Fath, B.D., Dean, C.A., Katzmair, H., 2015. Navigating the adaptive cycle: an approach to managing the resilience of social systems. Ecol. Soc. 20 (2).

Holling, C.S., 1986. The resilience of terrestrial ecosystems: local surprise and global change. Sustain. Dev. Biosph. 14, 292–317.

Jennings, F.D., 2000, January. The Role of NSF in "Big" Ocean Science: 1950-1980. In: 50 Years of Ocean Discovery: National Science Foundation 1950-2000. National Academies Press, Washington, D.C., p. 276

Ledford, H., 2015. Team science. Nature 525 (7569), 308.

Knauss, J.A., 2000, January. The emergence of the National Science Foundation as a supporter of ocean sciences in the United States. In: 50 Years of Ocean Discovery: National Science Foundation 1950–2000. National Academies Press, Washington, D.C., p. 276

Kubany, S.K., 2008. Musings on communications within the ocean research community. Oceanography 21 (3), 26–37.

Mazur, A., Boyko, E., 1981. Large-scale ocean research projects: what makes them succeed or fail? Soc. Stud. Sci. 11 (4), 425–449.

McKendall, V.J., 1996. Factors Facilitating Interorganizational Collaboration. Doctoral dissertation, University of Minnesota.

National Research Council, 1999. Overcoming Barriers to Collaborative Research: Report of a Workshop. National Academy Press.

Smith, R.W., 1989. The Space Telescope. Cambridge University Press, New York, p. 469.

Steele, J.H., 1983. Institutional and educational challenges. In: Oceanography. Springer-Verlag, New York, NY, pp. 377–380.

CHAPTER 1

The Bering Sea Project

Francis K. Wiese
Stantec Consulting Services, Inc., Anchorage, AK, United States

Introduction

The Bering Sea is home to a rich variety of biological resources, including at least 450 species of fishes, crustaceans, and mollusks, more than 36 million seabirds from 35 species, as well as 25 species of whales, seals, sea lions, and other marine mammals, and the world's most extensive eelgrass beds. Diverse and highly productive, this seasonally ice-covered sea has historically provided about 40% of the total US commercial fish catch with an annual value exceeding $3 billion (US dollars), and contains the world's largest sockeye salmon fishery. Critically, it also provides three-quarters of the subsistence harvest that supports 55,000 Alaska Natives and others living in more than 30 coastal communities (Wiseman et al., 2009). Many of these communities have existed around the Bering Sea for centuries with important cultural links to this dynamic marine ecosystem.

The timing and extent of the seasonal ice play an essential role in the productivity and community structure of this marine ecosystem. The early 2000s saw unprecedented heating in the southern Bering Sea (Stabeno et al., 2007) and climate scientists predict a major reduction in ice cover in coming decades (Stroeve et al., 2005; Zhang et al., 2010), with serious ecological and economic consequences intensifying concern over the future (Hunt et al., 2002; Mueter et al., 2006). To better understand and predict the large-scale ecological changes that could have major economic and cultural implications in the Bering Sea and elsewhere, the National Science Foundation (NSF) and the North Pacific Research Board (NPRB) joined forces to implement an extensive end-to-end marine ecosystem study.

Background

In early 2002, a group led by George Hunt from the University of Washington submitted a draft manuscript to NSF that described a potential large-scale oceanographic program focused on the impacts of climate change on the eastern Bering Sea ecosystem. As a result, NSF funded a series of international planning workshops, which eventually led to the development of the Bering Ecosystem Study (BEST) Science Plan (ARCUS, 2004) and Implementation Plan (ARCUS, 2005), and a series of Bering Sea focused, NSF funded, projects conducting fieldwork in 2007 (Hunt, 2012).

Meanwhile, the NPRB was finalizing its overall Science Plan (NPRB, 2005). NPRB was established by Congress in 1997 and first organized in 2001. The mission of the NPRB is to improve the understanding of marine ecosystems of Alaska and provide information to support effective management decisions on the sustainable use of Alaska's abundant marine resources science planning. Activities commenced in early 2002 and have since played a significant role in supporting marine research off Alaska. Guided by their Science Plan and input from the National Research Council (NRC, 2005) that emphasized the importance of large-scale integrated studies of the marine ecosystems, NPRB drafted an implementation plan for their first Integrated Ecosystem Research Program focused on the Bering Sea (BSIERP) in 2006. In doing so, it benefited from input through the Bering Sea Ecosystem Interagency Working Group (BIAW) established in 2005 in recognition that efforts to understand the role of sea ice in the Bering Sea ecosystem would require an interagency coordinated effort employing the unique capabilities of each contributing agency. Members of the working group included NPRB, NSF (through BEST), the US Geological Survey, the US Fish and Wildlife Service, the Alaska Ocean Observing System, the National Oceanic and Atmospheric Administration (NOAA) Alaska Fisheries Science Center, and Pacific Marine Environmental Laboratory, the University of Alaska Fairbanks, and the US Arctic Research Commission (Sigler 2006). Rather than going it alone, however, Clarence Pautzke, NPRB Executive Director, invited Bill Wiseman from the NSF Office of Polar Programs in charge of BEST, to join the NPRB Board meeting in September of 2006 to discuss a potential collaboration. Recognizing that this proposed collaboration and leveraging of resources would allow for a much more comprehensive ecosystem study than if each organization were to pursue something similar on its own, the Board unanimously endorsed the establishment of a partnership with NSF and approved $14 million (US dollars) to be dedicated to such an effort.

Building the partnership

Building a coordinated program between two organizations with distinct research and funding cultures was not a trivial task (Wiseman et al., 2009).

It was accomplished through the goodwill and support of the administrative and scientific leadership in both organizations, and the relationships, vision, and commitment of the individuals involved, while both recognizing the aforementioned rapid biophysical changes taking place in the Bering Sea. Different partnership models were explored, from concentrating funds and issuing a joint call to separate requests for proposals but coordinate fieldwork and annual meetings. NPRB's vision was a winner-take-all integrated team approach with specifically funded objectives and necessary review by the NPRB Science Panel and approval by the Board of Directors. NSF's model was about meritorious individual proposals contributing to the overall goals of the BEST Science Plan, reviewed by an independent panel of experts and recommended for funding to the Director of the Office of Polar Programs by the designated program officer.

After many meetings throughout the fall of 2006, the result was an administrative integration between NSF and NPRB described in a signed joint program management plan at the end of October. It specified a coordinated call for, and review of, proposals, and laid out key programmatic elements, including the scientific purpose of the program, each organization's funding commitments, the geographic scope of the programs, the division of emphasis on ecosystem components, review and selection of proposals, the establishment and maintenance of an integrated scientific team, the planning and coordination of field and modeling activities, as well as data collection, sharing and archival activities, and mechanisms for annual program adjustments if needed (NPRB, 2010). Importantly, this agreement also included a contingency plan clause whereby both parties agreed that in case they might fail to agree on a fully integrated program, each organization would retain the option to fund proposals of its choosing, including, if appropriate, not funding any proposals.

Scientifically, the partnership was centered around improving our understanding of how the highly productive marine ecosystem of the Bering Sea may respond to climate change, particularly as mediated through changes in seasonal sea ice cover. It envisioned a vertically integrated program that provides for end-to-end coverage of the Bering Sea ecosystem from atmospheric forcing and physical oceanography up through humans and communities, with the attendant economic and social impacts of a changing marine ecosystem. To align with institutional missions, and create synergies but avoid overlap, it was decided that NSF would provide support for projects focused on physical and biological oceanography up to and including macro-zooplankton and benthic infauna, as well as social science projects focused on relationships between a changing marine environment and the communities residing around the Bering Sea. In turn, NPRB would provide support for projects focused on trophic levels starting at macro-zooplankton and benthic infauna up to and including humans, their communities, and social and economic impacts. In addition, associated modeling efforts would be supported in both programs with the intent to, once funded, merge these into one integrated modeling effort from climate to fisheries and economics. This latter effort would be supported by an Ecosystem Modeling Committee (EMC) established by NPRB in September of 2006.

Program implementation

In late October 2006, the NPRB released a call for BSIERP pre-proposals. These were received and reviewed by the NPRB Science Panel and Board by early December. Two proposals were invited to submit full proposals in mid-December, coincident with the NSF BEST-focused solicitation. Through a confidential process, the cognizant program officers from each organization solicited comments from the cognizant program officers of the other organization concerning drafts of their proposed solicitations to ensure consistency with the signed partnership agreement. In their solicitations, NPRB sought applications from multi-disciplinary, multi-institutional teams of scientists, with the anticipation that one team would carry out the study. In contrast, NSF sought applications from individual investigators or groups of investigators with the intent to develop a team from the successful applicants, anticipating funding between seven and fifteen projects. Though separate, both solicitations included statements that they were striving for a fully integrated and coordinated program between NSF and NRPB and that they would be sharing proposals, conducting joint reviews, and making their recommendations in consultation with each other. It also indicated that successful applicants would be expected to sign and comply with provisions of a project management plan that would be developed by the assembled team based on requirements identified by NSF and NPRB.

To support the integration of fieldwork, the NPRB invitations for full proposals included a reference to ship schedules identified in the NSF solicitation, encouraged applicants to leverage NSF ship time, and to carefully explain any additional ship/cruise requirements. The NSF solicitation, in turn, stated that "additional opportunities to collect field data may become available through coordination with the winning team from the NPRB competition". In addition, both solicitations included statements that neither anticipated funding new climate modeling studies. Instead, applicants were required to indicate which of the IPCC model outputs would be used in developing their models and assumptions about climate-driven impacts on the ecosystem. It was foreshadowed that funded principal investigators would eventually need to agree on a set of common climate scenarios and assumptions for both lower and upper trophic level studies and that modeling components of proposals received by both organizations would need to address model evaluation criteria developed by the EMC. This level of programmatic coordination at this early stage was a key achievement to support the eventual integration.

Proposals for both programs were due in the spring of 2007 and technical reviews were conducted in April. Each organization selected their ad-hoc mail reviewers for the proposals they received following each organization's normal review procedures and conflict-of interest rules, but including a review of all modeling proposals submitted to both organizations by the EMC. In mid-June, a joint science review panel met an NSF. The panel consisted of the NPRB science panel and science director (to discuss NPRB proposals) and experts appointed by NSF (to discuss NSF proposals). All panelists were designated as formal NSF panelists, signed the NSF conflict-of-interests forms, and abided by NSF confidentiality rules. All panelists were briefed on joint instructions concerning the goals of the partnership. While each panel sub-group evaluated proposals submitted to their respective programs, the others would listen to the discussion. After each group ranked their respective proposals according to their criteria (laid out in the respective solicitations and the program management plan), the combined panel discussed how well the highly ranked proposals fit into an integrated ecosystem study and provided advice to the two organizations concerning how to optimize such a study. Where two or more proposals were believed to be equally meritorious from a scientific perspective, the joint panel provided a final ranking based on their view of the societal importance, e.g., their importance to managers, subsistence hunters, and commercial fishermen, as well as their contribution in the overall integrated program.

In late June, the full NPRB, select Science Panel, and EMC members, and William Wiseman from NSF, met to decide which of the integrated proposals should be funded, determine how it aligned with the NSF selected proposals, and identify any gaps that may need to be addressed.

Recommendations principally followed the recommendations of the joint science review panel, and lead to the following main additional efforts: The Local and Traditional Knowledge groups from various proposals were to come together to form one integrated plan, and the different modeling components were to meet with the EMC to refine the scope, ensure adherence to the published modeling criteria, and ensure tight coordination with the field efforts.

After the Secretary of Commerce reviewed and approved the recommendations made by the NPRB, and the head for the NSF Office of Polar Programs approved recommendations made by the cognizant Program Officer, the integrated NSF-NPRB funded partnership that would become known as the Bering Sea Project was launched in July 2007.

In the NOAA led successful integrated proposal to the NPRB, major in-kind resources ($15 million approximately) of scientists, equipment, and ship time, were made available by NOAA, USFWS, and USCG. Combined with the final investment from NSF ($21 million approximately), and NPRB ($16 million), the total program investment was close to $52 million, supporting over 100 principal investigators and many dozens of post-doctoral associates and graduate students from 32 academic, federal, state, and private institutions across the US and Canada (Wiese et al., 2012). The 43 linked components focused on key species, processes, and selected coastal communities, combining ship and community-based fieldwork, laboratory analysis, modeling, data management, and education and outreach, forming a highly integrated program from climate to fisheries and people (Fig. 1.1), structured around five core hypotheses:

1. Climate-driven changes in the physical components that control the Bering Sea (e.g., temperature, wind, sea-ice, and currents) modify the availability and allocation of food for all species.
2. Climate and ocean conditions influencing water temperature, ocean currents, and ecological boundaries impact fish reproduction, survival, and distribution, the intensity of predator-prey relationships, and the location of zoogeographic provinces.
3. Warming temperatures and subsequent earlier spring sea-ice retreat result in later spring phytoplankton blooms, thereby leading to increased abundance of piscivorous fish (e.g., walleye pollock, Pacific cod, and arrowtooth flounder) and a food web controlled by predators.
4. Climate and ocean conditions influencing water temperature, ocean currents, and ecological boundaries affect the distribution, frequency,

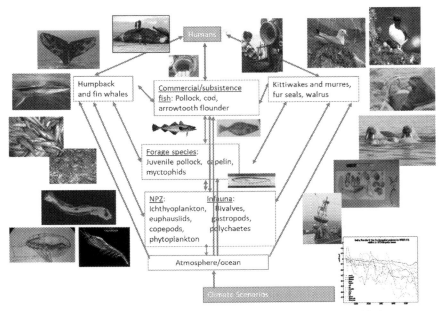

Fig. 1.1 Bering Sea Project schematic showing focal species and linkages.

and persistence of oceanographic fronts and other prey-concentrating features, and thus control the foraging success of marine birds and mammals.

5. Changes in climate and ocean conditions will affect the abundance and distribution of commercial fisheries and subsistence harvests.

Program management

The first meeting of the principal investigators for the joint Bering Sea Project was held in September 2007. Nearly 100 scientists and program management staff came together to get to know each other, review the scope and details of the program, work together toward achieving full program integration, identify gaps, discuss how to coordinate the field studies and cruise planning over the next 3 years, and work on a project management plan.

The Science Advisory Board (SAB) was established; a steering committee of six scientists (three supported by NSF and three by NPRB) elected from within the program, to work together with NPRB and NSF program staff to help build integration and move this ambitious effort toward successful outcomes. In the first months of the project, the SAB and the program staff worked on a Project Management Plan that was finalized and signed by all 100 PIs by the end of December 2007. The purpose of this plan was to identify and agree to binding provisions and protocols that promoted the development of a seamless team effort for scientific research. In effect, it laid out the expectations of NPRB and NSF for their funded principal investigators and included the overall program management structure (Fig. 1.2), the role of the SAB (including its membership and term limits), an anticipated PI meeting schedule, and expectations, field season planning and selection of chief scientists, communication and integration protocols among field programs and modelers, implementation and monitoring of data sharing protocols, coordination of education and outreach activities to achieve maximum synergies, coordination with other ongoing marine research programs and fisheries surveys in the Bering Sea, annual review and progress report requirements, reporting of research results and synthesis, and dispute resolution.

NPRB established and funded the Bering Sea Project office and hired a full-time program manager in June 2008. A detailed communication and outreach plan was written and approved, focused on sharing news from the field component among a variety of audiences, including coastal communities and national media. A dynamic website was created with constantly changing news for both the scientific community and the public (https://www.nprb.org/bering-sea-project/).

The first field season of this integrated program commenced in March 2008 with a cruise on the United States Coast Guard icebreaker Healy going from Dutch Harbor to St. Lawrence to study ice conditions, the

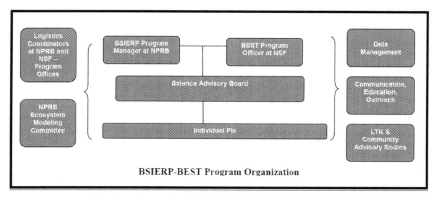

Fig. 1.2 Bering Sea Project management structure.

benthic prey fields, and walrus distribution. Between 2008 and 2010, field scientists went on to sample the eastern Bering Sea every month except December and working with community members in Savoonga, Emmonak, Togiak, St. Paul, Akutan, and several communities in the Nelson Island area. This interdisciplinary sampling, at higher frequency, across broader spatial scales than ever before attempted in this region, and the use of novel technologies, allowed the development of a more detailed and insightful description of the seasonal cycle of the ecosystem than previously possible. By the end, the pooled efforts of Bering Sea Project team members totaled over 24,000 safely accomplished person-days of fieldwork (more than 93 years for one person working full time year-round). Monthly SAB meetings and PI calls, annual PI meetings, regular EMC involvement, and much field-model integration and project component-specific discussions, guided and held the project together as a whole over 7 years. Scientists from across the project who initially identified as being from a specific institution and belonging to a specific project component became close colleagues and friends and became Bering Sea Project team members.

After several years of data analysis, laboratory studies, extensive modeling work, and synthesis, the Bering Sea Project culminated during 2013 and 2014. Continued funding from NPRB and NSF supported special synthesis efforts, development of project findings communication products, and enabled continuity in program management and the SAB until the end. During these final years, the SAB and project staff created a "Road Map" that mapped out objectives against published or planned research papers, enabling evaluation of the degree to which objectives initially laid out had been addressed. In cases where insufficient attention had been paid to a particular planned objective, the leadership team encouraged and facilitated stepped-up efforts. The Road Map was especially important to NPRB as they had paid special attention to proposed objectives during the evaluation phase, particularly as they pertained to fisheries management applications.

Program findings

Communication of project findings fell into four general categories with different emphases and focusing on different target audiences: (1) peer-reviewed publications; (2) scientific meetings and conferences; (3) general-interest events and materials; and (4) data archive and accessibility.

As a one-stop shopping, an integrated mechanism for communicating detailed scientific Bering Sea Project results, four "special issues" were published in Deep-Sea Research Part II (Ashjian et al. 2012, 2013, 2014, 2016). In total, the project spawned nearly 200 publications (and counting) across a wide range of journals. Many other papers, book chapters, theses, and dissertations have stemmed from the project. Project team members participated in dozens of scientific workshops, conferences, and symposia, delivering hundreds of conference presentations and many keynote talks. The capstone event was the Bering Sea Open Science Meeting, held in conjunction with the Ocean Sciences 2014 conference. With about 100 attendees from research and management backgrounds, the event met its goal to make research more accessible.

Using original text contributed by principal investigators, fifty two-page summaries called "Headlines", designed to translate individual project component findings into informative reports for the scientific community, industry, coastal communities, and the interested public, detailed notable project findings. From there, the 88-page Bering Sea "Magazine" was published in 2015 (NPRB, 2015). It serves as a resource for managers, policymakers, residents, resource users, and other stakeholders to understand and interpret Bering Sea Project findings.

A comprehensive data management support strategy for the Bering Sea Project engaged the science team early on to determine their requirements, establish priorities based on available resources, and establish clear metadata standards. NSF and NPRB supported the development and maintenance of the Bering Sea Project Data Archive at the National Center for Atmospheric Research's Earth Observing Laboratory (EOL), the single source for all data from this collaborative effort (http://beringsea.eol.ucar.edu).

This comprehensive, openly accessible database houses all the metadata and data collected during the Bering Sea Project, including a GIS tool for displaying detailed data and information collected during the local and traditional knowledge project on Nelson Island. The data and metadata can be perused through a search tool or displayed in tables by cruise, subject category, or investigator's name. With 360 datasets contained in the data archive, it is part of a powerful legacy of the Bering Sea Project.

Sustainability

The Bering Sea Project was a huge team effort sustained over seven years. Key to this sustained effort laid in the structural set up of the project, the continued hands-on involvement by the funding agencies, and project leadership. The program and scientific structures (Fig. 1.1 and 1.2) developed early on in 2006/7 laid out the interdependencies of the 43 components, a roadmap, and the necessary tools toward a successful integrated program. Initially, this forced new connections between people and disciplines that had not previously worked together, but over time, these relationships became self-sustaining, and relationships and common scientific products developed. The fact that NSF and NPRB stayed intimately involved throughout this process helped immensely, as it allowed for adaptive management and additional support were needed to keep the integration and scientific synthesis going, thereby keeping the team together from start to end. Finally, the SAB, whose membership changed little over the seven years, was key to this sustained effort. They were financially supported to meet monthly and kept the 43 components co-ordinated, energized, and on target vis-à-vis the project objectives. The relationships that developed and deepened over this time created the trust, a team spirit, and commitment, that eventually went beyond just meeting funding obligations.

Lessons learned

From a science management perspective, implementing and sustaining large-scale interdisciplinary research projects is an experiment in itself. This project was the first of its kind at this scale in this region and bringing together NPRB and NSF to spearhead this effort. As such we examine what we learned from this experiment so we can improve on it. In reflection of this sever-year long experiment, some of the lessons are as follows:

- *Scientific Framing*: The Bering Sea Project benefited greatly from the scientific frameworks laid out in the BEST Science Plan through the Oscillating Control Hypothesis postulated by Hunt et al. (2012) and in the BSIERP Implementation Plan and its integrated ecosystem vision developed by NPRB in 2006. The successful NOAA-led proposal capitalized on these frameworks to develop their guiding hypotheses. In combination, all this integrated scientific framing helped keep the program focused and on track toward new scientific insights, and allowed for the creation of applied results that benefited fisheries and ecosystem management in the eastern Bering Sea.

- *Partnership Building*: Having institutional champions that trust each other and are willing to push their internal systems for the good of the partnership is key for success. Formalizing the partnership in a written and signed agreement (in this case the NPRB-NSF Management Plan) helped to crystallize expectations, define commitments, and provide transparency to the scientific community responding to this opportunity.
- *Project Implementation*: Even though money was not mixed, integration between the NSF and NPRB funded teams was achieved. Key for this were the coordinated and cross-referenced RFPs and building synthesis into the project in its design, rather than being an afterthought. NSF funded individual projects with an expectation they contribute to a broader program with people unknown at that point, whereas NPRB funded an integrated team expected to form during the proposal process. This led to some NSF-funded PIs initially being more concerned about their funded project rather than the greater goal, but most of that went away through all the integration efforts and meetings described above. On the NPRB side, a winner-take-all approach for such a large project meant that true competition was tricky because scientific capacity at that scale is limited (evident by the small number of proposals received). It also meant that the overall goal drove the team, an approach that may not always result in "the best of class" for all project components.
- *Project Management*: Formalizing project management approaches and expectations in a team lead project management plan agreed to and signed by all involved was a key pillar throughout the project. The management structure (Fig. 1.2) laid out the leadership and having the SAB and the program officers manage the program hands-on over seven years to meet the integrated project applied goals was a cornerstone of its success.
- *Modeling*: The end-to-end modeling effort was a major component and an ambitious goal of the project. Effective use of end-to-end modeling requires that the models developed for a system are indeed integrated across climate drivers, lower trophic levels, fish population dynamics, and fisheries and their management (Punt et al., 2016). Although easily conceived conceptually and benefiting greatly from the guidance and continued involvement by the EMC, this was a major undertaking that needed additional leadership, resources, and time to be successful. The continued integration activities facilitated strong new connections

between modelers and observationalists, and led to joint modeling and field-data analyses well beyond the scope of the original program, contributing to direct advice to managers in making decisions regarding this ecosystem (Link, 2016).
- *Data and Data Management*: The Bering Sea Project benefited greatly from the availability of existing regional long-term monitoring data, especially the Bering Sea moorings and fisheries surveys operated by NOAA. Much context and ecosystem understanding would have been lost without those sustained efforts that allowed data analysis and modeling going back several decades. Undoubtedly, the metadata and data catalog left behind will be one of the program legacies. Having a central clearinghouse for all the project data to facilitate collaboration and synthesis proved key, but it also cost more and took longer than initially anticipated.
- *Communication*: For a project of this size, complexity, and importance to the region, clear communication within the team and to interested parties was key. Whereas there is never any doubt in these scientific endeavors that presentations at conferences will be given and scientific papers will be published, products that summarize relevant project findings to a broader audience, including local communities and policymakers, are less frequent. The outreach and communication developed early on specified different target audiences and associated communication vehicles that got implemented as the project progressed, was key, and included participation in local events (e.g. Bering Sea Days on the island of St. Paul, Alaska), interviews in local newspapers and radio programs, the development of brochures describing the project in general terms, and the creation of the project "Headlines".
- *Trust and Team-building*: Inter-disciplinary research is ultimately a social process. People need to get along and trust each other to work together, share information, and collaborate on publications, conference presentations, and other communication products. This takes time, and the program structure with its interdisciplinary science cruises, community visits, and its frequent meetings and interactions, facilitated this process.

Conclusions

To better understand and predict the large-scale ecological changes that could have major economic and cultural implications in the Bering Sea and elsewhere, nearly 100 principal investigators from scientific disciplines spanning climate, oceanography, fish and fisheries, seabirds, marine mammals, economics, anthropology, and ethnography, joined forces in the Bering Sea Project. Throughout this complex, seven-year integrated study of the Bering Sea, these principal investigators—together with their many colleagues, community members, technicians, students, ship officers and crews, and other field and laboratory teammates—covered everything from climate, physics to fish, subsistence, and management.

Humans are an integral part of the Bering Sea marine ecosystem in terms of commercial fisheries, a powerful economic engine for local fishing communities, the state of Alaska, and the nation, food-security for coastal communities, and ocean health. With this in mind, the Bering Sea Project explicitly included economic (Haney and Pfeiffer, 2013), ethnographic (Fienup-Riordan et al., 2013), subsistence (Fall et al., 2013), and local and traditional knowledge (Huntington et al., 2013) components, allowing the broader program to integrate these important socio-ecological ecosystem components with the natural sciences to draw connections and comparisons between these different ways of knowing and understanding the Bering Sea. Given the applied nature of its goals, the project had ramifications for the ongoing monitoring and management of the Bering Sea, with the Alaska Fisheries Science Center continuing the development of the integrated fisheries ecosystem model and the fieldwork needed to feed it. The aim is to operationalize the model to provide Ecosystem-based Fisheries Management (EBFM) advice on an ongoing basis (Link, 2016).

The success of the Bering Sea Project might be judged by its rich scientific data and publication legacy, its contribution to ongoing and future sustainable fisheries management, and by highlighting the societal importance of healthy marine ecosystems. Perhaps more intangible, but equally important, is how this strong partnership will be a model for forging new linkages between organizations, between researchers, between researchers and communities, as well as between physical, biological, and economic models, for investigating key marine ecosystem processes. What we know already is that this integrated approach brought forward a new generation of young researches interested in ecosystem science, and has changed how marine ecosystem science is done in Alaska today. Its successes and the many lessons learned detailed above have already influenced other projects including the Gulf of Alaska Project (https://www.nprb.org/gulf-of-alaska-project/), the Arctic Integrated Ecosystem Research Program (https://www.nprb.org/arctic-program/), the Marine Arctic Ecosystem Study (MARES) described in Chapter 5 in this book, and an integrated ecosystem approach to the Gulf of Maine (http://gulfofmaine.org/public/gulf-of-maine-council-on-the-marine-environment/programs/).

Acknowledgment

I am thankful to Clarence Pautzke to give me the initial opportunity to create and work on this project and partnership. Many people eventually contributed to its success and I would like to especially highlight George Hunt who had the vision to develop the BEST Science Plan that would become the complementary piece to BSIERP, Bill Wiseman at NSF who had the vision and drive to change the rules of the game, Tom van Pelt who was hired a couple years in as Program Manager and supported the vision to its completion, Nora Deans who as Coomunictions and Outreach Director had the product vision for multiple audiences, the Science Advisory Board (Mike Sigler, Carin Ashjian, Mike Lomas, Rodger Harvey, Jeff Napp, and Phyllis Stabeno) who provided key leadership throughout the entire process, and the over 100 other experts from communities, academia, government, and the private sector, without whom none of this would have been possible.

References

ARCUS, 2004. Bering Ecosystem Study (BEST) Science Plan. Fairbanks, AK: Arctic Research Consortium of the U.S, p. 82. https://archive.arcus.org/bering/reports/index.html.

ARCUS, 2005. Bering Ecosystem Study Program (BEST) Implementation Plan. Fairbanks, AK: Arctic Research Consortium of the U.S, p. 43. https://archive.arcus.org/bering/reports/index.html.

Ashjian, C.J., Harvey, H.R., Lomas, M.W., Napp, J.M., Sigler, M.F., Stabeno, P.J., van Pelt, T.I., 2012. Understanding ecosystem processes in the eastern Bering Sea III. Deep-sea Res. Part II 65—70.

Ashjian, C.J., Harvey, H.R., Lomas, M.W., Napp, J.M., Sigler, M.F., Stabeno, P.J., van Pelt, T.I., 2013. Understanding ecosystem processes in the eastern Bering Sea II. Deep-Sea Res. Part II 94.

Ashjian, C.J., Harvey, H.R., Lomas, M.W., Napp, J.M., Sigler, M.F., Stabeno, P.J., van Pelt, T.I., 2014. Understanding ecosystem processes in the eastern Bering Sea III. Deep-Sea Res. Part II 109.

Ashjian, C.J., Harvey, H.R., Lomas, M.W., Napp, J.M., Sigler, M.F., Stabeno, P.J., van Pelt, T.I., 2016. Understanding ecosystem processes in the eastern Bering Sea IV. Deep-Sea Res. Part II 134.

Fall, J.A., Braem, N.S., Brown, C.L., Hutchinson-Scarbrough, L.B., Koster, D.S., Krieg, T.M., 2013. Continuity and change in subsistence harvests in five Bering Sea communities: Akutan, Emmonak, Savoonga, St. Paul, and Togiak. Deep Sea Res. Part II Top. Stud. Oceanogr. 94, 274—291.

Fienup-Riordan, A., Brown, C., Braem, N.M., 2013. The value of ethnography in times of change: the story of Emmonak. Deep Sea Res. Part II Top. Stud. Oceanogr. 94, 301—311.

Haynie, A.C., Pfeiffer, L., 2013. Climatic and economic drivers of the Bering Sea walleye pollock (*Theragra chalcogramma*) fishery: implications for the future. Can. J. Fish. Aquat. Sci. 70 (6), 841—853.

Hunt Jr., G.L., 2012. Interagency Cooperation in Support of Bering Sea Ecosystem Science. https://www.arcus.org/witness-the-arctic/2012/3/article/19456.

Hunt Jr., G.L., Stabeno, P., Walters, G., Sinclair, E., Brodeur, R.D., Napp, J.M., Bond, N.A., 2002. Climate change and control of the southeastern Bering Sea pelagic ecosystem. Deep Sea Res. Part II Top. Stud. Oceanogr. 49 (26), 5821—5853.

Huntington, H.P., Braem, N.M., Brown, C.L., Hunn, E., Krieg, T.M., Lestenkof, P., Noongwook, G., Sepez, J., Sigler, M.F., Wiese, F.K., Zavadil, P., 2013. Local and traditional knowledge regarding the Bering Sea ecosystem: selected results from five indigenous communities. Deep Sea Res. Part II Top. Stud. Oceanogr. 94, 323—332.

Link, J.S., 2016. The importance of understanding ecosystem processes in the Eastern Bering Sea. Deep Sea Res. Part II Top. Stud. Oceanogr. 134, 424—425.

Mueter, F.J., Ladd, C., Palmer, M.C., Norcross, B.L., 2006. Bottom-up and top-down controls of walleye pollock (*Theragra chalcogramma*) on the Eastern Bering Sea shelf. Prog. Oceanogr. 68 (2—4), 152—183.

NPRB (North Pacific Research Board), 2015. In: Van Pelt, T.I. (Ed.), The Bering Sea Project: Understanding Ecosystem Processes in the Bering Sea. North Pacific Research Board, Anchorage, AK, p. 88. ISBN 978-0-692-54864-6. https://www.nprb.org/bering-sea-project/communications-outreach/.

NPRB (North Pacific Research Board), 2010. NPRB-NSF management plan for a study of the Bering Sea ecosystem. In: The Foundational Years 2001-2008. North Pacific Research Board, Anchorage, AK, pp. 267—274, 288 p. ISBN 978-0-9772670-1-9. https://www.nprb.org/assets/uploads/files/General_NPRB/nprb_reports/NPRB_Foundational_Years.pdf.

NPRB (North Pacific Research Board), 2005. North Pacific Research Board Science Plan. North Pacific Research Board, Anchorage, AK, p. 198. https://www.nprb.org/assets/uploads/files/General_NPRB/nprb_reports/NPRB_Science_Plan_2005.pdf.

National Research Council (NRC), 2005. Final Comments on the Science Plan for the North Pacific Research Board. The National Academies Press, Washington, DC. https://doi.org/10.17226/11235.

Punt, A.E., Ortiz, I., Aydin, K.Y., Hunt Jr., G.L., Wiese, F.K., 2016. End-to-end modeling as part of an integrated research program in the Bering Sea. Deep Sea Res. Part II Top. Stud. Oceanogr. 134, 413—423.

Sigler, M., 2006. Loss of Sea Ice Workshop. AFSC Quarterly Research Reports, April—June 2006. https://archive.afsc.noaa.gov/Quarterly/amj2006/divrptsHEPR1.htm.

Stroeve, J.C., Serreze, M.C., Fetterer, F., Arbetter, T., Meier, W., Maslanik, J., Knowles, K., 2005. Tracking the Arctic's shrinking ice cover: another extreme September minimum in 2004. Geophys. Res. Lett. 32 (4).

Stabeno, P.J., Bond, N.A., Salo, S.A., 2007. On the recent warming of the southeastern Bering Sea shelf. Deep Sea Res. Part II Top. Stud. Oceanogr. 54 (23—26), 2599—2618.

Wiese, F.K., Wiseman, W.J., Van Pelt, T.I., 2012. Bering sea linkages. Deep-Sea Res. Part II (65—70), 2—5.

Wiseman, W.J., Jeffries, M.O., Pautzke, C., Wiese, F., 2009. Two US programs during IPY. In: Influence of Climate Change on the Changing Arctic and Sub-arctic Conditions. Springer, Dordrecht, pp. 221—232.

Zhang, J., Woodgate, R., Moritz, R., 2010. Sea ice response to atmospheric and oceanic forcing in the Bering Sea. J. Phys. Oceanogr. 40 (8), 1729—1747.

CHAPTER 2

Belmont Forum partnerships

Erica L. Key
Belmont Forum Secretariat, Montevideo, Uruguay

Introduction

The Belmont Forum is a global partnership of funders committed to supporting transdisciplinary, transnational approaches to global environmental change. Founded in 2009, the Forum has grown its membership to include funding organizations on 6 continents representing scientific interests in 55 countries. In addition, the Forum's activities are supported by seven core partner institutions whose main function is not funding, but whose scientific programming, scope, or coordination role is aligned with Belmont Forum. Future Earth (FE) acts as a boundary organization to the Forum, providing connectivity to experts, stakeholders, and a constellation of coordinators and actors addressing global environmental change. Members contribute annually to the Secretariat, which is the primary implementing arm of the Forum. The Secretariat ensures that the actions of the members are instituted, including the delivery of international funding opportunities.

The vision that frames Belmont Forum activities is the Belmont Forum Challenge: *understanding, mitigating and adapting to global environmental change*. This challenge is periodically reviewed to ensure that it remains relevant and addresses a critical need that would benefit from transdisciplinary, transnational approaches. The current challenge was adopted by the members at the 2016 plenary meeting in Doha, Qatar.

Transdisciplinarity has never been formally defined in Belmont Forum governance documents but has come to be identified in a Belmont context as a co-developed, co-implemented activity that threads together different knowledge types to create an understanding that is more than the sum of its disciplinary parts. In the Forum, those disciplines are typically listed as natural (including physical) science, social sciences and humanities, and stakeholder knowledge. This foundational expertise can be enhanced by collaboration with other specialties, including but not limited to health sciences, engineering, communications, and computer and information sciences.

Transnational engagement adds geographic and cultural dimensions to transdisciplinary activity. The external benefit of this transboundary collaboration is the real opportunity to explore scalability and compare the application of informed approaches to global environmental change in different locales. The internal added value is the sharing of local knowledge and perspectives that can improve sensitivity and awareness of different ideologies and priorities amongst collaborators. Taken together, transdisciplinary and transnational activities have the potential to affect transformation, which is the underlying core of the Belmont Forum Challenge.

Partnership processes

The main instrument of delivery on the Belmont Forum Challenge is the Collaborative Research Action (CRA). The CRA is an international funding opportunity that develops from a theme proposed by a Belmont Forum member or Future Earth, the latter providing a bottom-up thematic proposal to the Forum informed by its global community. The theme should be relevant to the Challenge and must resonate with at least three funders in the membership to move into a scoping process. As a benefit of their membership, only members can propose themes (with the exception of the one bottom-up proposal from Future Earth) and vote on the theme proposals during the annual Belmont Forum plenary meeting.

Scoping a joint call

The CRA theme proposals that are accepted at plenary are then scheduled to transition to scoping. The activity is typically led by the organization that submitted the theme proposal, with support from the Belmont Forum Secretariat. CRA scoping engages experts, stakeholders, and a broad spectrum of resource providers to identify either a focal nexus or an envelope of interest around the theme. Scoping events are intended to be as inclusive as possible, viewing the theme through a system lens that incorporates input from a multitude of actors, including but not limited to academics, natural resource managers, government scientists and representatives, indigenous organizations, NGOs, and relevant knowledge holders and funders from all sectors. No bounds, geographic or otherwise, are imposed on participation in the open discussion. Involvement in the scoping process is voluntary and in no way influences later funding decisions should eligible participants choose to answer the CRA call for proposals. Similarly, funders who join the scoping discussion are not obligated to partner and provide resources in the final call announcement.

The end result of this exercise is a current, informed view of the theme. The information is then passed through the lens of the Belmont Forum Challenge and shared with the funding and resource community to identify synergies with planned investments that could be developed into partnership annexes.

This part of the scoping process is a closed negotiation amongst funders to ensure there is no conflict of interest or unfair advantage given to experts or stakeholders that provided early input on the theme. Scoping does not have a defined length or required number of engagements but instead allows the interest, immediacy, and alignment of funding programs and milestones to set the pace and frequency of meetings. The Forum works to engage with a variety of funding and resource providers to enable support for a broad geography and knowledge base to respond to the call. As the interested funder coalition starts to form becoming a Group of Program Coordinators (GPC), they elect an administrative lead from amongst themselves. This administrative lead, called the Thematic Program Office (TPO), consists of program contacts from one of more of the committed partners on the CRA; at least one of these has to be a Belmont Forum member. The TPO helps advance the development process and coordinates input on the call text and implementation plan. The Secretariat can assist the TPO to ensure that the final delivery is compliant with the Belmont Forum Challenge and procedures. Numerous support guides are also available on YouTube and the Belmont Forum resource page to advise interested TPOs and GPC partners.

Interested funders can be as engaged or as hands-off as they see fit during scoping. While some organizations can only partner if they have been involved from the outset and see relevant language in the call text, others may be able to partner at the end of the development process after the text has been formulated by other engaged organizations. It is their option, and the flexibility allows different funding cultures and workflows to partner on a CRA.

Funding a joint call

The Belmont Forum Secretariat assists the scoping process by engaging broadly with resource providers to increase the potential for inclusive support of the theme. Since its first Collaborative Research Action was launched in 2012, the Forum has partnered and engaged with resource providers on 6 continents, building relationships through BF-sponsored info days, guest invitations to its plenary meetings, connections through its members' international partnership offices, or other collaborative fora, such as the Global Research Council, the Heads of International Research Organizations, the Science Granting Councils Initiative, the Transformations Finance Forum, or the UN-Science-Policy-Business Forum. The Belmont Forum also participates in regional roundtables and on relevant committees that bring further exposure and foster familiarity with global environmental change resource providers.

Since the Forum solicits transdisciplinary, transnational approaches, the available project support must reflect a similar breadth. Forum governance allows both monetary and in-kind contributions to projects and partnerships with resource providers from all sectors. In-kind support can take many forms, including but not limited to high-performance computing time, cloud credits, laboratory space, access to collaborative platforms, and collaboration with already-funded experts. Not all resources come from members but instead include contributions from a combination of interested members and external partners (e.g., non-member resource providers) who value global, aligned activity around a relevant theme.

The Belmont Forum funding model is a virtual pot — monies and resources are not commingled and remain under the control of the providing organization. The organizations represented by the Group of Program Cooordinators committing resources to the call outline their interests in the theme, the nature of support, and eligibility rules for that support, as well as specific restrictions or emphases required for that support. This information is compiled into an annex that each partner prepares to join the call. The annex also features program contact details since only they have the legal authority to weigh in on the eligibility of a proposing institution or project theme relevance of a given proposal. Annexes may be submitted at any time before the (pre-)proposal deadline, even after the call has been launched. This is to allow a bit more flexibility to better align programming and allow time for different approval processes within partner organizations.

Designing a joint call

The call implementation development has a great deal of flexibility. The resource partners work together to identify:
- the number of phases in the call,
- whether the focus will be theme-forward or process-forward — for example focusing on climate predictability versus building climate adaptation capacity,
- any training or workshop opportunities to be offered before or during the ingress window,

- sub-themes or parallel tracks of funded activity, including networking, synthesis, research,
- special inclusion of required knowledge beyond the foundational transdisciplinary definition,
- length of the ingress window,
- and additional merit review beyond the core Belmont Forum criteria.

To ensure that the review process produces sufficient content to support award recommendations, it is important to survey the partner requirements during the development process and co-create a joint review program that meets all needs. Typically the organization that has the most comprehensive minimum requirement will set the bar for the review implementation.

It is recognized that some potential partners do not follow a merit review process, instead relying on board member votes or individual selection of awards. The Belmont Forum is developing alternative pathways to engage with these partners so that collaboration is possible without requiring a shift in decision culture. For example, it is possible for all partners to connect funded activity "after the fact", given the agreement of involved funders and project participants. These ad hoc arrangements do require more flexibility and understanding on all parts but are a useful partnership tool when programming timelines or review processes cannot be aligned. The Forum is also exploring specific support mechanisms, such as for CRA program coordination, early career researchers, or stakeholder consultation to leverage resources from organizations that are unable to commit to the joint merit review process.

Core templates are available for all implementation documents, including but not limited to conflict of interest forms, ingress templates, reviewer letters, panel of expert criteria, and data and digital object management plan guidance. The Thematic Program Office completes the documents needed to implement the call as outlined by the Group of Program Coordinators. Delivery of these documents together with the call text and annexes to the Secretariat are the final steps toward call launch on the Belmont Forum grant operations (BFgo.org) website. As a last collaborative step, any partners joining a Belmont Forum CRA for the first time are asked to sign a Memorandum of Understanding (MOU), ensuring they will follow the process as outlined in the call documents. At this time, the Belmont Forum is a global partnership without any legal authority, and so the MOU is essentially a good faith document between partners to implement the procedure they have co-designed.

Implementing the partnerships

Co-creation of a joint international call for proposals with a coalition of willing funders is only the first step. It is at CRA launch that the partnership moves into a public space, drawing wide attention to each organization's contribution. All funding partners are called upon to participate in the processing of proposals, expert review, award recommendations, and synchronized support. Given the global scope of ingress and the varied management styles of resource partners, it is a truly cross-cultural experience.

Call launch and review

Once the Group of Program Coordinators is ready to launch the call, all public documentation is made available on the belmontforum.org website and the Belmont Forum Grant Operations (BFgo.org) website. The communication burden is shared jointly by the Secretariat and all CRA partners to ensure that the opportunity is known to eligible proposers, using tools such as town hall meetings, exhibition events, social media, list-serves, web posts, and dear colleague letters to distribute information about the call. BFgo opens to receive submissions for that theme using the template(s) and deadline(s) agreed upon by the GPC.

Webinars are typically held to offer an opportunity for interested proposers to ask questions and better familiarize themselves with the Belmont Forum submission procedure. Proposers are encouraged to view the many online guides and tutorials on the Belmont Forum YouTube Channel to walk them through various aspects of the proposal process. Specific questions about call requirements are directed to info@belmontforum.org or for system ingress questions, help@bfgo.org. Due to conflict of interest guidelines, Belmont Forum does not provide a matchmaking service to connect interested proposers, but it can provide an unmoderated platform, such as a LinkedIn or WhatsApp Group, for potential applicants to come together. This particular function of proposal team development is an area where boundary organizations like Future Earth and partners like the International Science Council (ISC), Mountains Research Initiative (MRI), SysTem for Analysis, Research and Training (START), the InterAcademy Partnership (IAP), the International Institute for Applied Systems Analysis (IIASA), and the Group on Earth Observations (GEO) play a strong role, enabling new teams to form through their outward-facing engagement programs and network of networks. The new annual Sustainability Research and Innovation Congress, a joint vision of Belmont Forum and

Future Earth, will have a strong emphasis on connecting transdisciplinary (TD) practitioners in inclusive activity to grow community and awareness of TD-relevant funding opportunities.

Effectively what this workflow (Fig. 2.1) has done is simplify the submission process for an international collaborative activity, allowing one proposal to be submitted for joint review by multiple resource providers who have committed funds or in-kind contributions in a single, collaborative call. Interested proposers are encouraged to review the call text and associated annexes to understand not only the frame but available resources and eligibility.

Each proposal submitted must not only address the call theme and scope but be a co-developed and co-implemented effort that weaves together a minimum of natural science, social science, and stakeholder knowledge (transdisciplinary), and it must draw on resources from a minimum of three annexes providing in the support tabe for that CRA (transnational). Some CRAs may have other explicit requirements to encourage participation by a specific community, such as early career researchers or data and information scientists, or the call text may specify a certain deliverable or approach, such

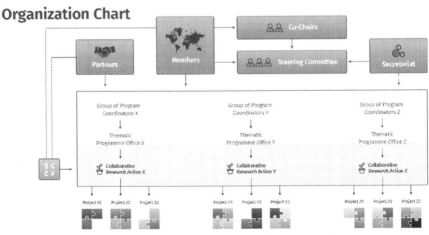

Fig. 2.1 This schematic details how multi-lateral funding partnerships are constructed in the Belmont Forum. Each project is supported by a minimum of three resource organizations who retain administrative and decisional control over their resources. Processes are synchronized and streamlined into a single ingress and review with aligned award dates to support international collaborative teams. The puzzle pieces represented in each project configuration could be interpreted either as the different knowledge types coming together, different countries collaborating, or the in-kind support and funds that were synchronized to advance the project. *Figure by LUX.*

as network development or synthesis. The TPO and all funding partners check submitted proposals in BFgo for the multi-lateral support requirement, conflicts of interest, compliance to the call criteria, and individual organizational eligibility requirements. Proposals that pass these administrative tests move forward for merit review consideration (Fig. 2.2). Submissions that do not meet the criteria are returned as ineligible for review.

The Belmont Forum uses a core set of review criteria for each CRA. Additional criteria can be added by the GPC for a given CRA, particularly if there are specific activities or collaboration guidelines outlined in the call text. Each eligible proposal is reviewed for quality and intellectual merit, stakeholder engagement and societal/broader impacts, interdisciplinarity, regional representation and quality of the consortium, appropriateness of resources, project management, and open data and digital objects management.

At a minimum, the TPO and GPC from each CRA convene a panel of experts that assess the consortia proposals against the Belmont Forum review criteria. Each resource partner supports up to 2 expert representatives on the panel, which may be an in-person or virtual event. The panel is led by a neutral chair with broad knowledge who is from a country not represented by the funding partnership. The chair's role is to ensure that the experts address the criteria in their evaluation in an unbiased way. The GPC identifies the panel chair, and the TPO supports his/her participation. If the panel is in-person, typically the TPO hosts the panel as well. The funders, themselves, are present to hear the discussion and to convene immediately

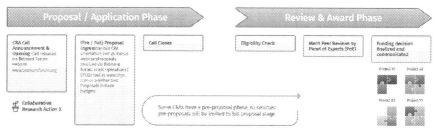

Fig. 2.2 The application, review, and award phases for Collaborative Research Actions (CRAs) follow a simple procedure that is agreed upon by all resource partners in advance of the CRA launch. Interested partners who do not subscribe to merit review processes but want to leverage the added value that these transdisciplinary, transnational activities provide can take advantage of new collaborative pathways beyond the CRA to work with the Belmont Forum. *Figure by LUX.*

post-panel, first with the panel chair, and then alone to recommend funding strategies. If the GPC decides they would like additional input to assist in their recommendations, the implementation plan for the CRA may also include ad hoc reviews, multi-phase panels, the requirement of letters of intent, or other such review structures. The end result should meet the needs of each organization on the GPC so that final recommendations pass muster with in-house policy offices; and so, the review process usually favors more input than less.

In most instances, funding is applied to eligible, meritorious projects until one or more organizations' resources are depleted. Sometimes funders can negotiate budgets, use annual funding instruments, or seek funding increases from their administrations to maximize the number of project participants they can support, while other funding partners may have the ability to apply resources internationally to address shortages. This post-panel discussion is a highlight of the Belmont Forum process, showcasing the ingenuity and goodwill of resource partners to expend the maximum amount of funding possible on meritorious multi-national projects.

Award management and evaluation

Once the list of recommended projects has been finalized by the GPC, each resource partner contacts those eligible for their support. At this time project, participants may be asked to upload their proposal into a proprietary award processing system so that the monies or resources can be disbursed. BFgo is many things — a GDPR-compliant proposal and review ingress portal, a post-award annual reporting and evaluation gateway, and an analytical tool — but it is not a financial management system. All award monies and resources are maintained and allocated by the resource partners directly. This commitment process is synchronized as much as possible with all other resource partners in the CRA to ensure that project teams can begin and end together without significant funding delays for one or more team members.

The award stage signals more than the beginning of individual projects. It triggers a cascade of events planned over the lifetime of the CRA to build transdisciplinary and open data and digital object management skills; support annual reporting to share lessons learned, best practices, outputs, and outcomes; create networks of transdisciplinary practitioners within and across CRA themes, and connect project deliverables with existing and developing coordination and implementation frameworks as well as new user and stakeholder groups. At a minimum, each CRA holds a (1) kick-off meeting to connect awarded teams, share expectations and milestones of the award cycle, and provide training resources for early implementation, (2) a midterm evaluation to explore challenges and opportunities of transdisciplinary approaches, initial outcomes, and connectivity with relevant frameworks, (3) and an end-term valorization that captures the total contribution and added value of the projects at that time.

This post-award activity is a collaborative undertaking implemented by the TPO, GPC, and Belmont Forum Secretariat together with interested partners and boundary organizations who are working in this thematic area. In more recent CRAs, there has been partial support of these events by coordination grants and contracts awarded through a variety of resourcing mechanisms, including support from European Research Area-Networks (ERA-Nets), single or coordinated multi-funder amendments to existing project awards, or direct funding by philanthropies, foundations, or aid and development agencies. There is potential that, with additional innovation, metadata tags, and funding of post-post award monitoring, additional activity could be conducted to capture the full scope of policy and practice contributions of the projects, recognizing that many impacts occur beyond the award lifetime.

Beginning in 2021, there is an emphasis for CRA milestones to take place at the Sustainability Research and Innovation (SRI) Congress that has been co-developed by Belmont Forum and Future Earth. The Congress fulfills a need for transdisciplinary, action-forward convening, providing a venue for Belmont Forum project teams to engage across themes and network with other sustainability actors and stakeholders. Unlike standalone or disciplinary engagements, SRI has programming relevant to all knowledge holders in the project teams and connectivity with transdisciplinary training, mentoring, innovation, policy, and funding. Just by holding CRA milestones at SRI, the added value of the activity is already amplified.

To inform these events as well as future activity of the Belmont Forum, all awarded project teams are expected to complete an annual report in BFgo. Since some funders do not require annual reporting or only information about their funded participants, and often those reports are confidential, the annual updates in BFgo represent the only comprehensive summary of the team's progress, challenges, and contributions. The BFgo report is required in addition to any funder reports specified in the award conditions.

Three templates exist to capture a combination of quantitative and qualitative data about the projects — an initial, interim, and final. Project lengths can differ, lasting from 1 to 5 years and requiring at a minimum, 1 annual report. For those beyond 3 years in length, additional interim reports are expected. Once multi-year reporting teams complete an initial template, information is carried forward to the next year's template for updating and addition. In this way, the annual progress of the entire team is captured while trying to reduce some of the reporting burdens.

The reporting templates themselves represent a collaborative approach toward collecting project and programmatic data to evaluate transdisciplinary efforts. A sampling of Belmont Forum member report styles and metrics derived from Forum governance documents were considered to develop baseline questions appropriate to transnational, transdisciplinary teams regardless of thematic focus. This core evaluation framework allows for project and programmatic intercomparisons across the portfolio. While no questions can be eliminated from this core, additional thematic-specific questions can be added through an annual reporting review process conducted with the GPC for each CRA.

The assessment of transformative research is in itself a research area. The Forum works with other organizations working at the science-society interface to advance metrics, indicators, and evaluation methods to best capture the contributions of the funded portfolio. An initial workshop was held and synthetic outputs developed toward potentially a suite of recommendations for evaluating transdisciplinary science. This is still an active area of inquiry and engagement for the Belmont Forum. The goal is a fully-realized evaluation program that captures local, immediate, and delayed, remote contributions of Belmont Forum's transformative projects and approaches.

Outcomes and lessons learned

The partnerships required to successfully operationalize the Belmont Forum platform are many — some targeted, others more comprehensive in their support and activity. The engagement process allows partners to choose their level and scope of commitment. Continuity is provided by the members and Secretariat that guide and implement the vision of the Forum.

Partnering

No matter the size of the program or budget, all resource providers face limitations in what they can support or have the capacity to administer. It is through the Belmont Forum partnership platform that these individual contributions can be aggregated to support something greater than the sum of its parts, and in doing so increase return on investment. The added value proposition must be there to warrant the additional time and process negotiation that the Forum requires. While much of the administrative burden is taken on by the TPO, participation in a CRA is not commitment-free.

Each organization has its views on added value, whether it is the advancement of knowledge, international partnerships, technology transfer, cross-sector governance, enhanced transparency, transdisciplinary training, network development, scholar exchange, preventative measures, informed management, or profitability. To reach the excitation level of collaborative partner, one must focus on helping organizations meet these goals through a flexible collaborative platform and with the minimal stress possible. Organizations are more likely to engage if the approach is familiar, the process open, and without significant policy or resource requirements. What Belmont Forum has done is create a simple bridge amongst existing systems and programs to streamline international and intersectoral collaboration while maintaining a global quality standard.

Attracting and onboarding new funding partners starts by building a trust relationship. By engaging and learning each partner's program priorities, funding philosophy, and decision process, the Forum is better able to create a working envelope that encompasses an area where partners are comfortable to collaborate. It helps to work through existing relationships, connecting regionally or sectorally through a local or familiar partner. The mechanics of actual engagement differs amongst organizations, but in-person meetings and an active interest in their outputs and leadership can make the difference. Even with current mobility issues, which have prompted the use of digital networking to continue partnership building, the online engagement is just an interim step until in-person meetings can be planned to concretize the relationship. Recently we have employed a virtual open house dialogue to allow more casual exchange amongst resource providers and coordinators, essentially trying to emulate the discussion one might have during coffee breaks at meetings. This informal chat is critical to building and maintaining programmatic relationships and is currently lacking in the efficient digital meeting structures we have adopted during quarantine.

Over the past 11 years that the Forum has been in operation, its reach has grown and welcomed new funding partners in nearly every CRA it has produced. Because of the different configuration of resource providers, each CRA has its own character and vision overlaid upon a common core of interest and commitment to the Challenge. In recent years, CRAs have fostered public-private cooperation and innovation, emphasized social sciences and humanities in the area of transformation and resilience, encouraged re-use of existing data sets and transdisciplinary outputs to generate new decision support tools and outcomes, enabled Low Middle Income Country participation through targeted funding from aid and development organizations and leverage of national capacity building commitments, integrated health sciences into all aspects of the portfolio, sought financial support for community stakeholder participation, enabled informed policy transformation, successfully funded projects that advanced both fundamental and applied sciences within a single consortium, and will now include specific training and research experiences for early career practitioners and those experiencing a mid-career shift toward transdisciplinary approaches. These accomplishments would not have been possible without the interest, dedication, and diversity of a global coalition of willing resource partners.

Resourcing

The scope of activity supported through the Belmont Forum touches on nearly every node in the financial resource ecosystem. The potential to bring together private sector, philanthropy, non-profit, advocacy, deep impact investment, non-governmental and public sectors support to bear on themes of shared interest is extremely powerful. Each support organization has its role to play in achieving commonly held goals — whether it is to train the next generation of leaders and decision-makers; grow the knowledge base; draft and implement informed policy; innovate, manufacture, and distribute; operationalize and manage; secure and protect; make accessible and transparent; transform and empower or some combination thereof. Each of these elements can be threaded together into a network of concerted or phased activity that helps achieve all of these mandates. By collaborating and co-creating an opportunity from the beginning, there is no hand-off mid-stream; the full cycle of endeavor is included in the framing to create more lasting opportunities and buy-in to the outputs and outcomes produced. Belmont Forum continues to work on achieving that potential with interested resource providers.

One way to ensure that the process is inviting to such a breadth of partners is to build a common vocabulary and address any perceived, whether real or otherwise, imbalances. In co-creating the call text, the verbiage should appeal to the range of potential proposing teams. Often it is a product of who are the main voices in the framing and may not achieve that total balance and inclusion. It is, in effect, a negotiation where funding availability from one source may be tied to the use of a particular word or phrase. Time also plays a factor in achieving the right tone. The hope is that at least the scope and intent of the call is clear and that targeted communication can be done to ensure that all knowledge holders feel relevant to the opportunity.

Some funding can come from organizations with broad remits where the representative involved in the call development may have a specific knowledge base. Where possible, Belmont Forum tries to engage program contacts from one or more relevant programs in these larger ministries and agencies to capture the range of interest. Certainly, the call text benefits from having natural science, social science and humanities, and other key expertise involved in its generation; but, it can also improve the resource contribution by ensuring that those voices were included from the very beginning and feel invested in the outcome.

Goals, current and future

One of the lingering challenges the Forum faces is the engagement of resources for open science. In many organizations the funds for data and information management are parsed in small amounts throughout programs; on the other end of the spectrum, they are centralized with sometimes a single staff member being responsible for all relevant investments. Since the Belmont Forum members adopted an open data policy and principles in 2015, it is critical to locate this support to implement open data and open access in all of its awarded projects. Many of the members and partners do not have organizational open data policies of their own, but through their experience with the Belmont Forum CRAs, they can "try-out" open science implementation and compliance checking. This policy has had a noticeable ripple effect with member and partner organizations that have since developed and adopted similar policies of their own. The next hurdle is to ensure that adequate resources are made available for projects to implement open science. Fully implementing an open data policy creates lasting added value through the re-use of generated data and digital objects toward new applications, advances careers through the valuation of data sets in promotional procedures and the use of citation and altmetrics in professional profiles, and moves us closer to achieving transparency goals for the distribution of taxpayer funds.

As well there is a need to find additional support for stakeholder participation in transdisciplinary projects. Since the term "stakeholder" is defined so broadly, there is not a single solution or class of support that would cover all possible participants. The funding portfolio has included community groups, government representatives at all levels, NGO representatives, unions and cooperatives, military officials, as well as individuals. When participation in the projects is not part of a stakeholder's paid job or livelihood, the resource providers should consider eligible support options. If not within the funding mandate of the CRA partners, it would be important to solicit additional organizations to join the Group of Program Coordinators to address this gap. Inclusion would likely bring additional funding as well as critical perspectives on appropriate and ethical engagement with stakeholders and an assurance that stakeholder needs are well considered in the review process.

The Belmont Forum remains committed to transdisciplinary, transnational approaches, and utilizing flexible partnering to aggregate the wealth of perspectives and resources that can be brought to bear on scientific challenges with societal relevance. These "grand challenges" truly benefit from an inclusive posture — developing a holistic view, co-designing opportunities, and embedding legacy through the involvement of stakeholder clients (and their relevant support base) — the total of which is still being determined through evolving transdisciplinary metrics and determinations of added value. At its most fundamental level, these partnership processes broaden our understanding of what is possible and globally important, and encourage respectful dialogue on how we can collaborate effectively to achieve a desired transformation.

Acknowledgment

The views and opinions expressed in this chapter are those of the author and do not necessarily reflect the official policy or position of her employer or any other agency or organization.

CHAPTER 3

The Nansen Legacy: pioneering research beyond the present ice edge of the Arctic Ocean

Paul Wassmann
Department of Arctic and Marine Biology, Faculty for Biosciences, Fisheries, and Economics, UiT - The Arctic University of Norway, Tromsø, Norway

Introduction

Changes in the Arctic have already had unprecedented impacts and consequences across a range of economic (Alvarez et al., 2020), environmental (National Academy of Sciences, 2007), societal (Stephen, 2018), and geopolitical (Tingstad, 2018) realities in the lower latitudes, most notably the rising sea level, increases in extreme weather, and substantial changes in international geopolitics. The Arctic and the northern oceans thus drive global-scale changes that further accelerate and amplify changes within the Arctic (IPCC, 2018). However, those changes, in turn, drive unprecedented changes affecting the rest of planet Earth, particularly the Northern Hemisphere (AMAP, 2017; Overland, 2020; Landrum and Holland, 2020).

Nowhere on the globe is climate warming more significant than in the Barents Sea region (Smedsrud et al., 2013; Lind et al., 2018), an inflow shelf which supports the greatest share of primary production in the Arctic Ocean (Sakshaug, 2004; Slagstad et al., 2015). The Barents Sea supports also one of the world's largest and best-managed fish stocks (jointly with Russia) (Nakken, 1998; Haug et al., 2017). From the Norwegian sector alone, 15 million fish meals a day are provided (equivalent to 8 high-quality fish meals a year for every European), in addition to increasing catches of snow crab. The physical conditions in the Barents Sea are characterized by contrasts between sea ice cover in winter and spring versus open water in summer and autumn. Observed changes include increases in seawater temperature, decrease in sea ice cover (e.g., Smedsrud et al., 2013), and change in ecosystem components (e.g., Fossheim et al., 2015) along a trend often described as "atlantification/borealization" (Polyakov et al., 2020). While the permanently ice-free southern part of the Barents Sea had been investigated for decades (e.g., Sakshaug et al., 1991; Eriksen et al., 2018), creating an ideal base for knowledge-based resource and ecosystem management, the hitherto ice-covered regions in Norway's northern territorial waters were significantly less investigated. To be able to manage resources and ecosystems also in this sector demands substantially more knowledge. This is the background for Norway's largest marine research project, the Nansen Legacy[1] with a total financial volume of about 81 million US$ over 6 years. What was the background and what were the applied procedures resulting in that Norway with only 5 million inhabitants invests so strongly into climate and ecosystem research in the seasonal ice zone of the Arctic Ocean?

Background

Knowledge-based resource and ecosystem management have long been a reality in the Norwegian sector of the Barents Sea. It started with the ProMare research project (1984—90). This was a national research project dedicated to investigating the ice-free water of the Barents Sea and the marginal ice zone with regard to physical, chemical, and biological oceanography, including fish, birds, and marine mammals. It was a field-intensive study that included (at that time) modern approaches, such as pelagic—benthic coupling, microzooplankton, nitrate- and silicate cycling, and physical—biological coupled ecosystem models. Organizationally, it was the first time two different research councils in Norway (Norges almenvitenskapelige forskningsråd (NAVF) and Norges fiskeriforskningsråd (NFFR) (before all research councils were united to the Research Council of Norway (RCN) in 1993) and the Ministry of Climate and the Environment fused their funds (about 25 million US$ in today's monetary value). Cooperation and coordination were the foundation of ProMare. In addition, the project was instigated by scientists, i.e., ProMare was a bottom-up initiative. All Universities and the federal Norwegian Institute of Marine Research were deeply involved, and a majority of today's senior Arctic Ocean scientists in Norway (many are already retired) were recruited for Arctic Ocean science through the project. Detailed information and reviews regarding the function of the Barents Sea ecosystem were published (e.g., Sakshaug et al., 1991). Of particular significance is Sakshaug et al. (2009), an updated book about the basic function and the different trophic levels of the Barents Sea ecosystem. Oceanographic conditions were

---

[1] https://arvenetternansen.com/.

summarized by Loeng (1991), and the pelagic ecosystem by Sakshaug et al. (1994). Oceanographic and biologic long-term trends have been addressed by Eriksen et al. (2017). The ProMare investigations were continued through regular annual investigations and a range of research projects. The variability and change of air—ice—ocean processes have been described by Smedsrud et al. (2013), summarizing the contribution of the Barents Sea to the Arctic climate system. The Norwegian CABANERA project focused on biological forcing of the carbon pump (Wassmann et al., 2006, 2008), and prospects for the Arctic Ocean seasonal ice zones with implications for the pelagic-benthic coupling, as outlined by Wassmann and Reigstad (2011). Detailed modeling primary and secondary production was carried out (e.g., Wassmann et al., 2010; Slagstad et al., 2011). The knowledge status for the Barents Sea is periodically updated in Norwegian management reports (e.g., Arneberg et al., 2020) and joint Norwegian/Russian Ecosystem Surveys (e.g., van der Meeren and Prozorkevich, 2019). Recently the remarkable role of subpolar advection for the over-all function of the Barents Sea has been highlighted (Vernet et al., 2019; Wassmann et al., 2019). The steadily increasing and updated knowledge of status, variability, and changes of physical, chemical, and biological systems that started during ProMare created the scientific basis for the Norwegian fishery management (for example, multispecies approaches) that paved the ground for the high standards that today's fishery management enjoys in the Barents Sea, with record-high biomass of cod.[2]

The Nansen Legacy is founded upon the expertise of ProMare and successive follow-up investigations. It applies the existing and tested organizational, logistical, and resource management expertise to the so far less investigated, but soon weakly ice-covered northern regions of the Barents Sea, the Svalbard archipelago, and adjacent Nansen Basin of the central Arctic Ocean (Fig. 3.1).

Norway's Arctic future?

For quite some time Norway looked, in concert with other Arctic coastal nations, to the less and less ice converged regions of the Arctic Ocean, contemplating about what the new conditions climate warming provided would entail for Norwegian fisheries, transportation to Asia, tourism, and oil and gas exploration: A result of this work resulted in the Norwegian

[2] http://www.mosj.no/en/fauna/marine/northeast-arctic-cod.html.

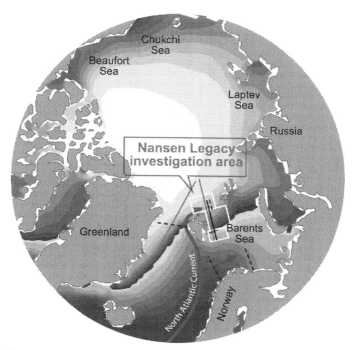

Fig. 3.1 The Arctic Ocean with its warm (red (dark gray in print version)) and blue water masses and permanently ice-covered (white), the seasonal ice zone regions (grades of blue (gray in print version)) and the North Atlantic Current. The research area of the Nansen Legacy project and transect lines are indicated. Illustration: M. Årthun, University of Bergen/Nansen Legacy.

Government's High North Strategy.[3] With the change from a center-left to a center-right government, the political effects of the conflicts regarding Crimea and eastern Ukraine, the recent drop in oil/gas price, and the expected decrease in future oil/gas demands changed earlier scenarios (High North strategy of 2011).[4] Some of the compelling optimism regarding resource extraction and economic possibilities (including the polemic upside-down turn of the map of Norway) and dedicated focus upon the Norwegian sector of the Arctic Ocean declined. However, much of Norway's economic interests in the High North are focused upon future fishing grounds for cod, halibut, shrimp, and snow crab, that experience not

---

[3] https://www.regjeringen.no/globalassets/upload/ud/vedlegg/strategien.pdf.
[4] https://www.regjeringen.no/globalassets/upload/ud/vedlegg/nordomradene/ud_nordomrodene_no_web.pdf.

only increasingly less but also highly variable ice cover. The solid base of today's knowledge-based resource and ecosystem management, the backbone of Norway's Arctic fisheries and environmental strategy, has thus to be transposed northwards into the hitherto ice-covered region of Norway's territorial waters. This sets up the backdrop for the Nansen Legacy. The knowledge foundation for future resource and ecosystem management is necessarily founded upon a science-based and multidisciplinary approach. The Nansen Legacy is a solid investment in Norway's sustainable use of its northernmost, Arctic waters.

First steps toward the Nansen Legacy project and goals

The original idea that resulted in the Nansen Legacy project was developed by a few scientists at the University of Oslo, the federal Norwegian Meteorological Institute, and the Norwegian Academy of Science and Letters in 2012. They saw Norway's increasing need for knowledge-based resource and ecosystem management in the ice-covered north and wondered how that challenge could be solved by any of the established research institutions and universities. Collaboration on the university or institute level seemed to be key to solve this grand national challenge.

The scientists approached UiT The Arctic University of Norway which, after probing the idea with other public institutions involved in marine Arctic exploration and research, applied for funding for a planning project. The project was ambitiously entitled Nansen Legacy right from the beginning, a name that connected the planned research to the novel and breathtaking endeavors of the Fritjof Nansen "Fram" expedition (1893—96). The application, signed by the principles of the universities and institutes [University Bergen (UiB), University of Oslo (UiO), the Norwegian University of Science and Technology (NTNU), the University Center on Svalbard (Unis), the Norwegian Polar Institute (NPI), the Norwegian Institute of Marine Research (IMR), and the Norwegian Meteorological Institute (met.no)] was submitted by UiT in February 2012. The application was sent inside a governmental funding scheme entitled SAK (samarbeid, arbeidsdeling og konsentrasjon, i.e., cooperation, division of labor, and focus) to the Norwegian Ministry of Education and Research (MER). The SAK principles created the backbone for the project. I will come back to the SAK principles toward the end because this funding scheme has bearing on the solution of many Arctic Ocean challenges.

The aim of the application was to receive financial support to develop a research proposal for the exploration of Norway's so far ice-covered waters that were only rudimentarily investigated. Under the leadership of UiT, a high-level Board with representatives from all institutions would be formed to plan the work and discuss the principal matters. The Board charter would also include the development of instructions for a Working Group with centrally placed senior scientists from all institutions and follow up on their work. The Working Group, in turn, would create a research plan to achieve its goals. As envisioned, the Board would have no authority to dedicate resources to the project. Rather every single institution willing to collaborate in the Nansen Legacy would have to decide about their type and level of involvement when it came to science, personnel, ship time, and other resources. The idea was that both the Board and the Working Group were supposed to launch the project for organizations that may be interested in supporting it financially, including the RCN, the EU, and industry. Also, the role of the project inside the frame of international cooperation in the Arctic Ocean was part of the mandate.

The vision was to gather the relevant climatic, oceanographic, and ecological competence of Norway for a coordinated research endeavor in the northern Barents Sea. The involved scientists would be challenged to develop innovative approaches related to the future knowledge and management demands regarding climate, environment, and ecosystem understanding for the High North. In early August 2012, MER decided to support the Nansen Legacy planning project with 290.000 US$. Shortly after, in a constituting meeting, the Board elected the author as chair of the Board.

Challenges and activities

The first and ultimate challenge of the Board was to develop the agreement for cooperation between eight institutions that, which so far and according to general rules, had been competitors. Should they, for once, cooperate, why, and how? The obvious reason to cooperate was quickly detected. Norway experienced a significant need for information on the extensive, ice-covered regions of the northern Barents Sea and adjacent central Arctic

Ocean. This demand for knowledge was so significant that no single institution had the resources and know-how to carry out the work alone in the foreseeable future. What were the alternatives?

Let us illustrate the challenge with an example from architecture. The might of a Greek temple depends upon the number and thickness of columns and its connecting architrave. The might and beauty of the "superstructure," the pediment where most of the elaborate sculptures and spectacular friezes are found, is based upon the number and might of the columns. Which in turn are raised upon a solid and step-formed basement. If one has great expectations for an extensive and splendid pediment, these have to be based upon fundamental knowledge (base), a multitude of contributors (columns), and their organizational connection (architrave). Cooperation and coordination connect institutions, and this creates the fundamental precondition for high expectations and anticipations. If that cannot be achieved (e.g., because of competition), the columns remain isolated structures that tower into an empty sky, almost asking to get connected with an architrave to raise the pediment.

How the potential cooperation should be organized and carried out was a great challenge that needed a wide range of Board meetings in which the philosophy and the principles of cooperation were discussed. What would be the benefit for the nation and for every single institution to join forces in the Nansen Legacy? Would this compromise the work that each institution wished to carry out? How should time on the new research ice breaker "Kronprins Haakon"[5] that was built during these years (Norway's first Arctic research vessel since the "Maud" was built in 1918) be used and how could the Nansen Legacy be constructed around this essential vessel for efficient polar research? Would the financial support for Nansen Legacy come exclusively from various funding agencies or would the participating institution contribute from their funds? At the end of the first year, the cooperative base and the funding principles had become a reality. The organization structure was simple. The Board, under the leadership of UiT The Arctic University of Norway, had one representative that represented his/her institution. Indeed, the eight institutions wished to lower their competitive shields and enter a phase of cooperation which was found beneficial for the goals of each institution. This was an oral agreement that was never fixed by a contract. Each partner was allowed to contribute, in cooperation with others, with their

---

[5] https://www.npolar.no/en/kronprins-haakon/.

specific fields and nobody had to give up expertise to create space for others. Below the Board was a Working Group (see below) with two scientific representatives from each institution Each partner institution promised to contribute 50% of their respective total costs inside the Nansen Legacy, while the rest had to come as fresh money.

An important principle of the Board meetings was that they were hold in turn at each of the partner institutions. This to underline that every partner institution was considered equally important and implied mobility for all partner institutions. To streamline tactical, political, and institutional aspects of the cooperation, joint meetings between the Board and the Working Group were organized en route to the evaluation of the research proposal. After about one year that was necessary to figure out the principles of the potential project, in early 2014, the Working Group was formed and through intensive and dedicated workshops it developed the research proposal of Nansen Legacy. To keep a good connection with the Board the chair of the Board participated in most of these workshops.

The Board had also meetings with various Norwegian partners such as the RCN, relevant ministries that would benefit from the project, and industry partners. Frequent meetings with the RCN were essential because RCN was supposed to have a central role in carrying out and quality control of the project. Three issues were controversial and needed time to solve. First, RCN felt uncomfortable that eight institutions had started to plan a large project without providing the research council the possibility to plan the work, on behalf of Norway. Bottom-up approaches were welcome, but preferentially through an assessment by RCN. Second, the formation of a national team including all important state-supported institutions for Arctic research implied that competition between research consortia became impossible. The competitive principle is strong in science and this frequently impedes united national teams. In the case of Nansen Legacy, the competitive principle between institutions had to be set aside for the benefit of the national team. Third, some commercial partners that deal with Arctic research complained that the significant funding provided to Nansen Legacy gave state-supported institutions a competitive benefit. At the final stage of Nansen Legacy, commercial partners such as Akvaplan niva[6] and NERSC[7] were included.

[6] https://www.akvaplan.niva.no/.
[7] https://www.nersc.no/.

Presenting the plan to industry and international partners

None of the industry partners such as prominent oil and gas companies on the Norwegian continental shelf were interested in participating in and co-financing Nansen Legacy. This viewpoint found support in the notion that (a) the companies were heavily taxed by the Norwegian government, (b) they followed the environmental obligations imposed upon them, and (c) the general state of the Arctic environment was considered first of all a public obligation. The response by the ministries for fisheries, environment, oil and gas, and foreign affairs was positive, but none of them showed direct interest to support Nansen Legacy. It is not known if these ministries argued in support of the Nansen Legacy through the government.

The Board also made sure that much of the international Arctic Ocean expertise was informed about the Nansen Legacy. Specific talks were presented in Canada, China, Denmark, France, Germany, Japan, Korea, Russia, the United Kingdom, and the United States. The research and cooperation plans were also presented to the Nation Science Foundation and discussed during meetings with the US Interagency Arctic Research Policy Committee in Washington DC.

Quality control of the research plan

Early stages of the research plan were presented during dedicated symposia to purposely invited, international Arctic Ocean experts whose main role was to constructively criticize the plan and scrutinize the up-to-datedness of the project. Eventual weaknesses in science and organization were constantly on the minds of both the Working Group and the Board. To receive formal external review and feedback, the final research proposal[8] was sent in late 2015 to the National Academy of Sciences in Washington DC who appointed an evaluation committee including national experts in marine ecology. The criteria for the evaluation focused upon the quality of the (a) the research, (b) the consortium, (c) the project leader and co-leaders, (d) the principal investigators, (e) the organization of the project, (f) gender balance, (g) international collaboration, and (h) the project's potential/anticipated impact. In early 2016, the plan of action and management strategy were found to be sound, and the leader, co-leaders, and

---

[8] file:///Users/pwa000/Downloads/The%20Nansen%20Legacy-Proposal-1.pdf.

scientific staff were found outstanding. The evaluation committee unanimously endorsed the Nansen Legacy Project. Based upon this evaluation, the Nansen Legacy project was submitted to RCN and MER. RCN subjected the application for an additional evaluation with experts of their choice.

Funding the Nansen Legacy project

After RCN declared that they would fund a major portion of the Nansen Legacy in 2016 (about 20 million US$) and the partner institutions guaranteed to fund 50% of their own costs, MER finally designated the missing funds (20 million US$) from the federal budget toward the end of the year. Nansen Legacy had a preliminary start already in March 2017 (MER provided also 1 million US$ for this start-up project), began in full in 2018 and has a planned ending in 2023. The total costs are 81 million US$. It presents an impressive team of 140 scientists, of which more than 60 are PhD students and postdocs. The project is an attempt to create a novel and holistic Arctic research project. It should provide the necessary, integrated scientific knowledge on the rapidly changing marine climate and ecosystem of the ice-covered and thus inaccessible Barents Sea and north of Svalbard. Nansen Legacy is providing the new knowledge base that is required to facilitate a sustainable management of the northern Barents Sea and adjacent Arctic Basin through the 21st century. All data are taken care of in a publicly available data base. The research team includes interdisciplinary arctic marine expertise from physical, chemical, and biological oceanography, as well as meteorologist, geologists, modelers, and underwater robotic engineers. Jointly, they are investigating the past, present, and future climate and ecosystem of the northern Barents Sea. In total, the Nansen Legacy will spend about 360 days at sea, using the research icebreaker "Kronprins Haakon" as its main research platform. The ship-based sampling is complemented by the use of underwater robotics, year-round moored observing platforms, and satellite-based observations. Complementary model tools are used to integrate field-based observations, and to investigate the dynamics of the physical and biological components of the northern Barents Sea climate and ecosystem at present and in the future. Nansen Legacy has an active outreach office and policy. The project will produce a textbook that addresses both managers and students.

Recruitment for a sustainable Arctic future

While the Board planned the Nansen Legacy project and contributed to building up a balanced multidisciplinary team, it figured out that more than 60 of Norway's senior expertise in polar research was just retired or close to be so. Nobody had taken note of that and no particular recruitment steps were being made to substitute the out-phasing expertise. The Nansen Legacy saw this immediate need for recruitment and made sure that most of the fresh funds were dedicated to PhD students and postdocs. Particular emphasis was given to PhD students that can enjoy a range of multidisciplinary courses where the students meet many of their future colleagues, accompanied by established research teams. This is another attractive aspect of national research team. They can take care of nation-wide recruitment, support the necessary multidisciplinary understanding, and support rather a collective "we" than the traditional "I." After the end of the Nansen Legacy, we anticipate that Arctic research in Norway will enter a new phase. Like ProMare before the Nansen Legacy has already become a game-changer.

The significance of cooperation, division of labor, and focus for solving grand national and international research questions

Over time every nation-state meets major challenges and must solve them. To permanently solve these challenges often institutes or agencies are established. They deal, for example, with weather forecasts, fishery quotas, agriculture management, or environmental problems. The institutes or agencies act inside a domain defined by the state. Each institution makes sure that it is restrained to what the defined tasks are. In this manner, a nation-state gets the necessary advice on how to tackle challenges and manage the country in wise and sustainable manners. However, the "map" of challenges is continuous and changes over time, while the responsibilities of the agencies often remain static. If so, then significant knowledge and expertise gaps may appear on the "map" that are not covered by any agency or institute. New problems, such as climate change, may appear and require informed advice for resource managers and policy makers, advise that can only be partly provided by the existing institutes and agencies.

The government could tackle this challenge by amending or expanding the mission of all affected institutions to include the new problems. For demanding challenges such as those addressed by the Nansen Legacy and a reasonable small country like Norway none of the institutions would be able to expand sufficiently to tackle the task. Also, the extensive expertise of the universities, goldmines of Artic knowledge, has to be involved. The frame of reference of many institutions had to be changed or expanded simultaneously, accompanied by a cross-institutional cooperation agreement. These forms for cooperation are challenging because many institutions have to defend their interests and compete for funds against each other. An alternative would be to establish an additional agency but that would demand that the new responsibilities, in particular regarding potential overlap with other agencies, are defined and that one more, long-lasting and static organization is created. Over time the question arises how many inert and distinct agencies a nation-state wishes to run? Are there alternatives to how a state can get hold of necessary information, knowledge, and advice without founding new agencies or changing the mission of existing ones? In other words, are there other ways to solve grand national challenges that might otherwise remain un- or only partly answered? Can such challenges be solved in a transient and cost-efficient manner?

Here the principles of cooperation, division of labor, and focus, the base of the Nansen Legacy, come into play. The idea is that not only state-run institutes and agencies but also universities and commercial companies can cooperate, create sustainable alliances, and form national teams. To what degree will the potential conglomerate of such bodies be able to cover the grand questions a state may have? And how could one in concert cover the gaps that arise between the various bodies? To form such temporary national teams, that are abolished after the end of the mission, a set of rules have to be established. All participating partners have to be interested to *cooperate* in a voluntary and clearly defined manner. No institution should exclude the interests of others if important for solving the question. That implies that the work has to be carried out jointly: there must be a *division of labor*. Although one partner may be dominating a certain field of expertise, other partners cannot be excluded: one has to divide or share a job. The last principle is that of *focus*. Although the division of labor is mandatory inside a national team, every partner must have the possibility to focus on a certain specialty to develop increased excellence.

Based on this approach, national teams can be built that are temporary and only exists for the length of the defined project. This results in cost-efficiency and provides a mechanism by which the entire expertise of a state can be used to solve a defined problem in a reasonable period of time that no partner could solve alone. In this way, the investment by the state is not fixed permanently to a new agency, but is kept changeable to solve forthcoming, grand questions. The idea of national teams in science is rarely evaluated. In the case of Norway, the Nansen Legacy was the first national team that was formed. It seems on track to become a veritable success. Never before have so many worked altruistically for a grand research challenge, while simultaneously supporting the expertise and position of each institution. Unquestionably, the cooperation makes all to winners. Nation-states and governments may learn from the process of cooperation, division of labor and focus, and the formation of temporary national teams. A priori there are no plans to maintain the partnership beyond the time of the project. A successful national partnership should only be kept alive if further grand research question of national concern arises.

Regretfully it does not look like national research teams will find much support in the future. One of the main reasons is, also in Norway, that such a team necessarily will result in the exclusion of competition principle, which seems so entrenched that alternatives seem unwise and impossible. The principle of competition and resistance against a national team was so strong that RCN suggested to divide the Nansen Legacy project into several parts and submit these fractions for competitive evaluation. In other words, one wished to rather sacrifice a novel and holistic Arctic research project for the sake of the competition principle. This attitude, strongly supported by MER (that abandoned the cooperation, division of labor, and focus approach), misses an important point: the grand challenges of a nation can only be solved with a national team. The grand challenges of a nation can under no circumstances be addressed through competition between among institutions when the expertise of all key instructions and most of the manpower are needed to address them. The search for the best solutions for a nation should have a higher priority than political principles. The question is if organizations that fund science should consider an entire range of funding instruments instead of the simplest one: the one and only competitive principle. The relationship between competition and the effect of funding is inversely u-shaped. With increasing competition, the efficiency increases significantly, reaches a maximum before additional increases in competition decrease the national efficiency. The challenge is to find the maximum of the relationship.

"The scientific investigation of a rapidly changing northern environment leads to research questions of such intellectual, empirical and logistical complexity—and of such importance to the management of national resources and associated international obligations—that they can only be addressed properly through national and prioritized cooperation, with the highest scientific standards." This quote from the Nansen Legacy documents places the endeavor into perspective. The Arctic future of Norway has become the focus of a temporary research program, supported by the entire nation. The Nansen Legacy is a game-changer that smaller, less efficient, but competing projects will never achieve.

Lessons learned?

Let me first convey that my involvement in the realization of the Nansen Legacy belongs to the greatest successes of my science career. Under the leadership of Marit Reigstad and UiT The Arctic University of Norway, the project is well on its way, successful and lives up to expectations. In only 6 years the Nansen Legacy provides Norway's inescapable need for a knowledge-based foundation for a sustainable ecosystem and resource management in the hitherto ice-covered regions of the High North. The project strengthens Norway's sovereignty in these waters and the nations reputation for wise and balanced management. A new and imperative generation of multidisciplinary Arctic specialists get recruited. All that suggests that there should be a strong motivation to initiate future national team projects which are so extensive that only a national team can pursue them. In smaller nations national teams may be the only cost- and time-efficient approach to solve the grand questions that the nation may face.

The Nansen Legacy approach benefitted from a bottom-up approach. Without any political constrains it were the scientists themselves who made an analysis of what the nation needs. Scientists are more insightful in national research need questions than research councils or other institutions close to the government make us believe. Many premiums that stand close to the government are frequently too influenced by political guidelines and institutional interests to ensure a fundamental and balanced analysis. The grand questions of a nation that only a national team may answer adequately should be fundamentally defined and not be a priori confined by guidelines by the government and the bodies that finance research. To listen to free science and to involve the science community at large may be the best possible approach a nation may have to look into the future.

However, as long as governments and research councils have not decided how to finance national teams, that by definition cannot be subjected to competition, the pathway of the future teams is insecure and may be impossible. Nansen Legacy is ready to share the experience with other national teams, but it may remain the only fully national team as the official Norway hesitates to support bottom-up initiatives of the national team idea. Free initiatives, creative thinking, and the development of fresh ideas, far away from the restrictions by either research organizations or governments, will rejuvenate science and keep the danger of "sclerotization" at bay. Bottom-up approaches are not a goal in itself to identify research needs, but an additional instrument for the benefit of the science scene and an efficient manner to obtain inevitable results. That the Nansen Legacy solves Norway's inescapable need for basic knowledge to guarantee a knowledge-based basis to ensure a sustainable ecosystem and resource management in the hitherto ice-covered regions of the High North is indisputable. It is Norway's first and remain its only national team, unless the idea of national teams is accepted as a potential strategy for which the necessary financial support mechanisms are created and sustained.

How will the future Arctic look like and how should we adapt to research efforts?

We can already now imagine how the principal features of a future Arctic Ocean may look like (e.g., see Spiridonov et al., 2011; Wassmann et al., 2015; Carmack et al., 2016; CAFF, 2017). It will have a cover of thin sea ice that in winter will cover the central basins and parts of the adjacent shelves. Because of decreased freshwater supply caused by less sea—ice volume melt and increasing low-pressure passage and strength, increased mixing will result in a shrinking of the rim of the seasonal ice zone. Most of the central Arctic Ocean will be free of sea ice cover in late summer and the remaining "last ice" will stay close to northern Greenland and the northeastern Arctic shores of Canada. The expanse of ice flora and fauna will decrease but decreasing ice thickness will result in more extensive ice alga blooms that compete with phytoplankton for nutrients. Increased warming and stratification on inflow shelves may result in a decrease in productivity. However, the exposure of the Arctic shelf breaks will result in strongly increased harvestable production along the Eurasian and eastern sections of the American shelf break (Slagstad et al., 2015). Through the ongoing borealization (e.g., Fossheim et al., 2015; Polyakov et al., 2020) the biodiversity will be negatively impacted, and Arctic species may get restrained to survival pockets, similar to previous warming periods. Fisheries will expand into the Arctic Ocean shelves, in particular the northern inflow shelves of the southern Chukchi and northern Barents and Kara Seas. However, the strong stratification and the low nutrient concentration in the central Arctic Ocean will probably prevent a sustainable pelagic fishery.

These trends are examples of some of the marine ecological changes in the Arctic Ocean of the decades to come for which we need exact knowledge. While these changes provide new economic possibilities that have to be explored, a sustainable use has to be our prime focus of every Arctic government. The Arctic Ocean, historically less accessible for humanity, accommodates specific ecosystem, high biodiversity, unspoiled wilderness, and charismatic animals such as whales, seals, birds, and the polar bear that mean a lot for humans. Our adaptation to climate change in the Arctic should encompass, at a minimum, adequate investigations of this "new ocean," the only new ocean humanity will ever get. Knowledge-based resource, ecosystem management, and a conservative, ecological definition of the widely misapplied term sustainability should govern our decisions. An important aspect of these endeavors are pan-Arctic science publications that inform the scientific community at large of what is known, and for this purpose conceptual models (Wassmann et al., 2020) have been developed to communicate essential results to managers and the general public (Wassmann, 2006; 2011, 2015; Wassmann et al., 2021). National and international cooperation appears essential to achieve results that are in support of Arctic Ocean ecosystems, Arctic people, and humanity at large. For successful cooperation, the principles of cooperation, division of labor, and focus could create a successful foundation in favor of pan-Arctic integration. The Nansen Legacy, conceptual endeavors, and the symposia for pan-Arctic integration are influenced by the SAK principles and are the contribution of UiT The Arctic University of Tromsø to the indispensable comprehension of the Arctic Ocean.

Every involved nation should seriously scrutinize their ways of funding science. We know now that a) the principal distinction between basic and applied science is false and b) that there is harmony, not dissonance, in making the transition from new ideas to a practical solution (Narayanamurti and Odumpsu, 2016). Many governments and founding organizations try to solve the traditional conflict between basic and applied science by strongly forcing research into an applied direction to enjoy rapid impacts on the economy. That will result in short-sighted progress and Pyrrhic victories. A wise and sustainable knowledge-based resource and ecosystem management for the hitherto ice-covered regions of the northern Barents Sea, for example, through the Nansen Legacy, supports applied science on the base of up-to-date basic science. It represents a modern and wise approach to sustain and consolidate Norway's landing of fish that in 2018 passed 2.2 billion US$ (430 US$ per capita).

Acknowledgments

The invitation and editorship of Guillermo Auad and Francis Wiese to contribute to this book is gratefully acknowledged. The impact that Ø. Hov, E. Elvevoll, A.-H. Hoel, B. Corell, and E. Carmack had upon my thinking and organization skills is gratefully acknowledged. This work was funded by the Research Council of Norway through the project The Nansen Legacy (RCN #276730). It is also a contribution to the UiT-based research group Arctic SIZE (http://site.uit.no/arcticsize/).

References

Alvarez, J., Yumashev, D., Whiteman, G., 2020. A framework for assessing the economic impacts of Arctic change. Ambio 49 (2), 407–418.

AMAP, 2017. Snow, Water, Ice and Permafrost in the Arctic (SWIPA), 2017. Arctic Monitoring and Assessment Programme (AMAP), Oslo, Norway. https://swipa.amap.no.

Arneberg, P., van der Meeren, G., Franzen, S., Vee, I. (Eds.), 2020. Status of the Environment in the Barents Sea. Report from the Advisory Group on Monitoring, pp. 1893–4536. Report 2020-13, 1154 pp. ISSN (In Norwegian).

CAFF, 2017. State of the Arctic Marine Biodiversity Report. Conservation of Arctic Flora and Fauna International Secretariat, Akureyri, Iceland, 978-9935-431-63-9.

Carmack, E.C., Yamamoto-Kawai, M., Haine, T.W., Bacon, S., Bluhm, B.A., Lique, C., Melling, H., Polyakov, I.V., Straneo, F., Timmermans, M.L., Williams, W.J., 2016. Freshwater and its role in the Arctic Marine System: sources, disposition, storage, export, and physical and biogeochemical consequences in the Arctic and global oceans. J. Geophys. Res. Biogeosci. 121 (3), 675–717.

Eriksen, E., Skjoldal, H.R., Gjøsæter, H., Primicerio, R., 2017. Spatial and temporal changes in the Barents Sea pelagic compartment during the recent warming. Prog. Oceanogr. 151, 206–226.

Eriksen, E., Gjøsæter, H., Prozorkevich, D., Shamray, E., Dolgov, A., Skern-Mauritzen, M., Stiansen, J.E., Kovalev, Y., Sunnanå, K., 2018. From single species surveys towards monitoring of the Barents Sea ecosystem. Prog. Oceanogr. 166, 4–14.

Fossheim, M., Primicerio, R., Johannesen, E., Ingvaldsen, R.B., Aschan, M.M., Dolgov, A.V., 2015. Recent warming leads to a rapid borealization of fish communities in the Arctic. Nat. Clim. Change 5 (7), 673–677.

Haug, T., Bogstad, B., Chierici, M., Gjøsæter, H., Hallfredsson, E.H., Høines, Å.S., Hoel, A.H., Ingvaldsen, R.B., Jørgensen, L.L., Knutsen, T., Loeng, H., 2017. Future harvest of living resources in the Arctic Ocean north of the Nordic and Barents Seas: a review of possibilities and constraints. Fish. Res. 188, 38–57.

IPCC, 2018. Summary for Policy Makers. Global Warming of of 1.5°C. IPCC Special Report, 2018.

Landrum, L., Holland, M.M., 2020. Extremes become routine in an emerging new Arctic. Nat. Clim. Change 10 (12), 1108–1115.

Lind, S., Ingvaldsen, R.B., Furevik, T., 2018. Arctic warming hotspot in the northern Barents Sea linked to declining sea-ice import. Nat. Clim. Change 8 (7), 634–639.

Loeng, H., 1991. Features of the physical oceanographic conditions of the Barents Sea. Polar Res. 10 (1), 5–18.

Nakken, O., 1998. Past, present and future exploitation and management of marine resources in the Barents Sea and adjacent areas. Fish. Res. 37 (1–3), 23–35.

Narayanamurti, V., Odumpsu, T., 2016. Cycles of Invention and Discovery: Rethinking the Endless Frontier. Harvard University Press, p. 170.

National Academy of Sciences, 2007. Arctic Environmental Change and Potential Challenges. https://www.nap.edu/read/11753/chapter/6.

Overland, J.E., 2020. Less climatic resilience in the Arctic. Weather Clim. Extremes 30, 100275.

Polyakov, I.V., Alkire, M., Bluhm, B., Brown, K., Carmack, E.C., Chierici, M., Danielson, S.L., Ellingsen, I.H., Ershova, E., Gardfeldt, K., Ingvaldsen, R.B., 2020. Borealization of the Arctic Ocean in Response to Anomalous Advection from Subarctic Seas.

Sakshaug, E., 2004. Primary and secondary production in the arctic seas. In: The Organic Carbon Cycle in the Arctic Ocean. Springer, Berlin, Heidelberg, pp. 57–81.

Sakshaug, E., Hopkins, C.C., Øritsland, N.A., 1991. Proceedings of the pro mare symposium on polar marine ecology, Trondheim, Norway, 12-16 May 1990. Polar Res. 10 (1), 1–4.

Sakshaug, E., Johnsen, G.H., Kovacs, K.M. (Eds.), 2009. Ecosystem Barents Sea. Tapir Academic Press.

Sakshaug, E., Bjørge, A., Gulliksen, B., Loeng, H., Mehlum, F., 1994. Structure, biomass distribution, and energetics of the pelagic ecosystem in the Barents Sea: a synopsis. Polar Biol. 14 (6), 405–411.

Slagstad, D., Ellingsen, I.H., Wassmann, P., 2011. Evaluating primary and secondary production in an Arctic Ocean void of summer sea ice: an experimental simulation approach. Prog. Oceanogr. 90 (1–4), 117–131.

Slagstad, D., Wassmann, P.F., Ellingsen, I., 2015. Physical constrains and productivity in the future Arctic Ocean. Front. Mar. Sci. 2, 85.

Smedsrud, L.H., Esau, I., Ingvaldsen, R.B., Eldevik, T., Haugan, P.M., Li, C., Lien, V.S., Olsen, A., Omar, A.M., Otterå, O.H., Risebrobakken, B., 2013. The role of the Barents Sea in the Arctic climate system. Rev. Geophys. 51 (3), 415–449.

Spiridonov, V.A., Gavrilo, M.V., Krasnova, E.D., Nikolaeva, N.G., 2011. Atlas of Marine and Coastal Biological Diversity of the Russian Arctic. WWF, Moscow.

Stephen, K., 2018. Societal impacts of a rapidly changing Arctic. Curr. Clim. Change Rep. 4 (3), 223–237.

Tingstad, A., 2018. Climate, Geopolitics, and Change in the Arctic. RAND Corporation Santa Monica.

van der Meeren, G.I., Prozorkevich, D., 2019. Survey Report from the Joint Norwegian/Russian Ecosystem Survey in the Barents Sea and Adjacent Waters. August-October 2018. IMR/PINRO Joint Report Series 2, 85 pp.

Vernet, M., Ellingsen, I.H., Seuthe, L., Slagstad, D., Cape, M.R., Matrai, P.A., 2019. Influence of phytoplankton advection on the productivity along the atlantic water inflow to the Arctic Ocean. Front. Mar. Sci. 6, 583.

Wassmann, P., 2006. Structure and function of contemporary food webs on Arctic shelves: an introduction. Prog. Oceanogr. 2 (71), 123–128.

Wassmann, P., 2011. Arctic marine ecosystems in an era of rapid climate change. Prog. Oceanogr. 90 (1), 1–17.

Wassmann, P., 2015. Overarching perspectives of contemporary and future ecosystems in the Arctic Ocean. Prog. Oceanogr. 139, 1–12.

Wassmann, P., Reigstad, M., Haug, T., Rudels, B., Wing Gabrielsen, G., Carroll, M., Hop, H., Falk-Petersen, S., Slagstad, D., Denisenko, S.G., Arashkevich, E., Pavlova, O., 2006. Food web and carbon flux in the Barents Sea. In Structure and function of contemporary food webs on Arctic shelves: a panarctic comparison. Prog. Oceanogr. 71, 232–287.

Wassmann, P., Carroll, J., Bellerby, R.J., 2008. Carbon flux and ecosystem feedback in the northern Barents Sea in an era of climate change. Deep-sea Res. Part 2. Top. Stud. Oceanogr. 55 (20–21).

Wassmann, P., Slagstad, D., Ellingsen, I., 2010. Primary production and climatic variability in the European sector of the Arctic Ocean prior to 2007: preliminary results. Polar Biol. 33 (12), 1641–1650.

Wassmann, P., Reigstad, M., 2011. Future Arctic Ocean seasonal ice zones and implications for pelagic-benthic coupling. Oceanography 24 (3), 220–231.

Wassmann, P., Slagstad, D., Ellingsen, I., 2019. Advection of Mesozooplankton into the Northern svalbard shelf region. Front. Mar. Sci. 6, 458.

Wassmann, P., Carmack, E.C., Bluhm, B.A., Duarte, C.M., Berge, J., Brown, K., Grebmeier, J.M., Holding, J., Kosobokova, K., Kwok, R., Matrai, P., 2020. Towards a unifying pan-arctic perspective: a conceptual modelling toolkit. Prog. Oceanogr. 102455.

Wassmann, P., Krause-Jensen, D., Bluhm, B., Janout, M., 2021. Towards a unifying pan-arctic perspective of the contemporary and future Arctic Ocean. Front. Mar. Sci. 8, 654. https://doi.org/10.3389/fmars.2021.678420.

CHAPTER 4

The Argo Program

Dean Roemmich[a], W. Stanley Wilson[b], W. John Gould[c],
W. Brechner Owens[d], Pierre-Yves Le Traon[e], Howard J. Freeland[f],
Brian A. King[c], Susan Wijffels[d], Philip J.H. Sutton[g] and
Nathalie Zilberman[a]

[a]Scripps Institution of Oceanography UCSD, San Diego, CA, United States; [b]National Oceanic and Atmospheric Administration, Washington D.C., United States; [c]National Oceanography Centre Southampton, Southampton, United Kingdom; [d]Woods Hole Oceanographic Institution, Falmouth, MA, United States; [e]Mercator-Ocean International, Toulouse, France and Ifremer, Plouzane, France; [f]Institute of Ocean Sciences, Fisheries and Oceans Canada, Sidney, BC, Canada; [g]National Institute of Water and Atmospheric Research, Wellington, New Zealand

Introduction

The Argo Program is the first of its kind — a systematic *in situ* global ocean observing network sustained for more than 20 years. Argo is accumulating critical data for basic research, climate forecasting and assessment, education, and operational oceanography. Over 2.3 million Argo temperature-salinity-pressure profiles have been collected, managed, and made available since 1999 (Fig. 4.1). Argo is built on a foundation of partnership. US Argo is a multi-institutional partnership, with major funding from NOAA, responsible for implementing the US national component of Argo.

The international Argo partnership coordinates more than 25 national Argo Programs (Fig. 4.2) that deploy Argo floats and report data to Argo's Global Data Assembly Centers (GDACs). The GDACs provide free and open data access in near real-time. Argo has been called "one of the scientific triumphs of the age[1]" by the New York Times and it is among the most collaborative efforts in the history of oceanography.

To understand Argo, we begin with its most direct roots in the World Ocean Circulation Experiment (WOCE), describing the genesis of the Argo partnership from these roots. This includes the formation of the international Argo Science Team (AST), the multi-institutional US Argo Program and its international counterparts, and the Argo data management system. Argo grew from a pilot program to become a central element of the Global Ocean Observing System (GOOS). During its first

---

[1] J. Gillis, New York Times, 2014: https://www.nytimes.com/2014/08/12/science/in-the-ocean-clues-to-change.html

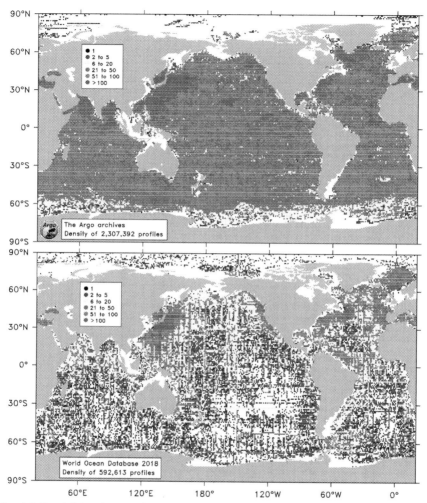

Fig. 4.1 Spatial density per 1° × 1° square, of all 2,307,392 Argo profiles, (Upper panel, 1999–9/2020) and all 592,613 non-Argo temperature-salinity-pressure profiles to depths greater than 1000 m (Lower panel, all years through 9/2020).

decade (1999–2009), Argo deployed a 3800-float global array that has been sustained at this level ever since while improving data quality, global spatial coverage, and data delivery. The international partnership that initiated Argo remains as its bedrock, even as Argo continues major growth in new directions.

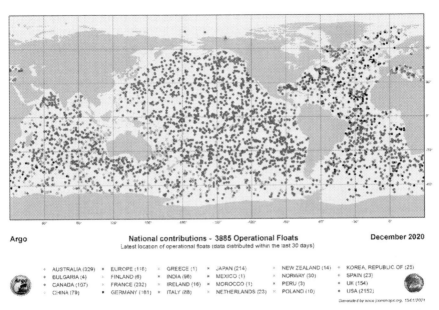

Fig. 4.2 Location of the 3885 operational Argo floats as of December 2020, color-coded to indicate Argo National Programs. *(Source: OceanOPS).*

Argo's roots in WOCE

In the 1990s, WOCE carried out an unprecedented global survey of the oceans, revealing important oceanic roles in the Earth's climate system (e.g., Stammer et al., 2003). Storage and transport of heat, mass, and freshwater were all addressed, based on oceanic property distributions and circulation. Later analyses would show that the oceans absorb more than 90% of the net heat gain (global warming) in the climate system (Rhein et al., 2013). This excess heat is distributed both horizontally and vertically, not only in the upper ocean but throughout the deep sea. Elements of WOCE included global hydrographic observations made by the international fleet of research vessels (RVs) and repeated upper ocean temperature profiling using eXpendable BathyThermograph (XBT) probes deployed from RVs and commercial ships. The WOCE global hydrographic "snapshot" was an enormous undertaking that took 7 years to complete. Above all, WOCE highlighted the need for continuing systematic observations needed to track the trajectory and patterns of climate variability and change (Siedler et al., 2001).

In addition to demonstrating the need for sustained global observations, WOCE also produced a revolutionary technology advance that created the capability for a global ocean observing system. The WOCE profiling float (Davis et al., 2001), descended from a lineage of neutrally buoyant floats (Gould, 2005), combined a high-pressure pumping system for buoyancy adjustment with a Conductivity-Temperature-Depth (CTD) sensor package plus satellite navigation and communications. Today's profiling float, powered by lithium batteries and communicating via low Earth orbit satellite networks, routinely cycles hundreds of times between the sea surface and 2000 m, and recently to 6000 m. Each cycle returns a high quality CTD profile and a velocity estimate based on the float trajectory. The profiling float is revolutionary because it enables climate-quality data to be obtained anywhere in the global ocean without an RV. The removal of this vessel constraint means that the physical state of the oceans can be measured globally, simultaneously, and repeatedly. The seasonal and hemispheric biases of historical data collection are eliminated in Argo. Fig. 4.1 shows the density in 1-degree squares of the 600,000 non-Argo CTD profiles[2] deeper than 1000 m in the World Ocean Database for all years (Boyer et al., 2013). The figure similarly shows the density of the 2.3 million CTD profiles obtained since 1999 by the Argo Program. The revolutionary advance in spatial coverage and observation density made by the profiling float is evident.

WOCE involved a global consortium of countries and institutions coordinated by an international steering team, with development and sharing of common procedures to process and distribute the global data set. To facilitate scientific analyses of these measurements, a data management system was created that included a common format for exchanging data and the development and sharing of processing and quality control procedures.

Thus WOCE provided the scientific motivation, the requisite profiling float technology, and prototypes for program organization and data management that underpin the Argo Program. Additional motivation came from the Tropical Ocean Global Atmosphere Project (TOGA, McPhaden et al., 1998), which showed, through tropical and extra-tropical observations, patterns and predictability in El Niño/Southern Oscillation variability. Together, WOCE and TOGA revealed the broad spectrum of

---

[2] This includes modern profiles obtained by electronic CTDs, mainly since the 1980's, and older profiles made by collecting water samples and lowering deep-sea thermometers.

ocean–atmosphere interaction and variability that defines the intra-seasonal to multi-decadal elements of the Earth's climate system. The success of WOCE, TOGA, and other early global-scale observing projects led to planning for the GOOS, applying the new capabilities for systematic sampling of the oceans. A critical element of the GOOS would be the Argo profiling float array (Gould, 2003).

The formation of Argo's multi-national partnership

Following the conclusion of the WOCE observational period in 1997, and given the nexus of climate science motivation and new technical capability, a group of scientists involved in the international CLIVAR[3] research program advocated the development of a global array of profiling floats. At the same time, the establishment of an international Global Ocean Data Assimilation Experiment (GODAE, Smith and Lefebvre, 1997; Le Traon et al., 2001) was under discussion. The GODAE objective was ambitious: to use new satellite observing capabilities and advances in modeling and data assimilation to forecast the ocean on a global scale and in real-time. For GODAE to become a reality, global real-time measurements of the interior ocean were needed to complement satellite measurements. Argo would satisfy this requirement, and the establishment of the Argo network was encouraged and supported enthusiastically by the GODAE and CLIVAR teams.

In mid-1998 the group of scientists was endorsed by the CLIVAR Upper Ocean Panel and GODAE to weigh the feasibility and consider the design of a global array of profiling floats. This group became the Argo Science Team[4] (AST), later renamed the Argo Steering Team. For the past 22 years, the AST has been the coordinating body for Argo's multi-national partnership. In 1998 the AST produced a consensus design for the global array (AST, 1998). The design of Argo was based on previous *in situ* ocean observations, including the WOCE datasets, and on related satellite observations including sea surface height (SSH). The name Argo was chosen to emphasize strong complementarity with the Jason series of satellite

---

[3] CLIVAR — Climate and Ocean: Variability, Predictability and Change
[4] Original members of the AST were D Roemmich (USA, chair), O Boebel (Germany), Y Desaubies (France), H Freeland (Canada), B King (UK), P-Y Le Traon (France), R Molinari (USA), WB Owens (USA), S Riser (USA), U Send (Germany), K Takeuchi (Japan), and S Wijffels (Australia).

altimeters. The Argo design called for a global, uniformly distributed array of 3300 free-drifting profiling floats, spaced about every 3-degrees of latitude and longitude. Each float would provide a CTD profile every 10—15 days from the sea surface to 2000 m.

The AST was composed of ocean scientists at academic institutions and government laboratories. Most of the world's experience with profiling floats was held by AST members. The AST provided the scientific vision and technical ability to design and implement Argo, but it did not command resources. A key step for Argo was to entrain the interest and commitment of the many international agencies responsible for sustained ocean observations in their respective nations. In the US, NOAA Administrator D. J. Baker and Deputy Chief Scientist W. S. Wilson were convinced, through discussions and early AST meetings, of Argo's great potential for fundamental ocean/climate science as well as its feasibility. The efforts of Baker and Wilson articulating their support for Argo, were critical for mobilizing partner agencies in many nations for the purpose of implementing Argo. Subsequently, Argo developed in the US and internationally as a partnership at both scientific and agency levels. Implementation of these partnerships at the national level was carried out by national Argo Programs such as US Argo, linked together by the AST and supported by national agencies. Further, given the potential benefits of Argo observations to the rapidly-growing operational oceanography effort, as well as to climate and oceanographic research, the Argo partnerships expanded to encompass both of these communities.

The newly established National Oceanographic Partnership Program (NOPP) served as a key avenue for expediting the Argo partnership within the US The NOPP had been created in 1997 to facilitate collaborative undertakings. When Baker made Argo NOAA's top priority new start in 2000, the Office of Naval Research, working through the NOPP, provided the initial funding to launch US Argo a year sooner, in 1999. The successful 1999 NOPP proposal was titled "An Integrated System for Real-Time CTD Profiling Float Data on Basin Scales (ARGO)" and funded the partners of the US Argo Float Consortium[5]. Some international partners had similarly strong ocean-related interagency collaboration while others relied on single agencies. In France, a high-level ocean committee developed the

---

[5] The U.S. Argo Float Consortium: Scripps Institution of Oceanography, University of Washington, Woods Hole Oceanographic Institution, and the NOAA laboratories PMEL and AOML, plus commercial partners.

Coriolis partnership and the Mercator Ocean forecasting center as the French contributions to Argo and GODAE, respectively. Argo implementation planning meetings for the Atlantic, Pacific, and Indian Oceans, organized by the newly formed national Argo Programs, were held to initiate or strengthen connections across agencies of the participating countries.

The approach by NOAA's Baker and Wilson to counterpart agencies in other nations helped initiate political visibility and support, and hence funding, for Argo in its early years as it was building coverage. As Argo grew toward completion and observations were increasingly collected on a broad scale, the importance of such early political support gradually diminished. Continuing support was increasingly justified by the impact the observations were having by meeting both research and operational needs. Close coordination between the parallel scientific and agency partnerships, facilitated by AST and other Argo meetings, allowed Argo to be rolled out in a sustained and growing fashion consistent with the developing technology and national capacities.

NOAA initially committed to supporting one-third of the array through the US Argo Program. A few years later, recognizing that most of the global capacity and expertise in profiling floats was in the US, the NOAA commitment was increased to half of the global array. Major international partners who joined at the outset and committed to providing floats included Australia, Canada, France, Germany, Japan, and the UK Soon many more nations, including India, China, the Republic of Korea, and others, joined the Argo partnership. Later, Europe established a legal structure to organize and federate European contributions to Argo (Le Traon, 2013).

Argo strives for uniformity and consistency across national programs through the AST's consensus-driven decision-making. All float-providing national programs have AST representation, and all have agreed that sustaining global coverage and highest data quality are Argo's top priorities. Most, but not all, national Argo Programs are supported by research, rather than operational, funding. This presents both advantages, such as prioritizing highest quality data, and challenges, such as continuity in funding, for long-term stability and growth of Argo.

Implementation of the global Argo array

Although the potential of the profiling float was evident by 1997, several aspects of technical readiness needed improvement before global deployment could be undertaken and sustained. These advances would lower the cost of Argo and would increase its spatial and depth coverage, characteristics that define the feasibility and sustainability of Argo and determine the success of Argo's international partnership.

(i) Float lifetime was short in the first years of Argo, and increasing float lifetimes was critical. For all Argo floats deployed in 2000−02, only 31% remained operational after 3 years. This increased to 63% after 3 years for deployments in 2003−05, and 72% after 3 years for deployments in 2016 (93% for US Argo).

(ii) Float buoyancy adjustment, and hence depth range, was marginal for global deployment, requiring tedious precision ballasting that depended on deployment location. Most newer float models are capable of 0−2000 m depth range anywhere in the ocean.

(iii) Argo would require an order of magnitude more profiling floats than the few hundred that were deployed in WOCE. The challenge of producing and deploying floats in sufficient numbers to sustain Argo, and with minimal early failures, has been met.

(iv) The low power CTDs developed especially for profiling floats had to be deployed and monitored to assess and improve their accuracy and stability.

A defining characteristic of Argo is the ongoing improvement of floats, sensors, and other Argo technologies, and these improvements have been critical for Argo's sustainability. Improvements in energy efficiency and battery technology make today's floats capable of 10-year battery lifetimes. Satellite networks now provide bi-directional communications, with greater bandwidth for increased data volume, reduced energy use, and a major reduction of time spent on the hazardous sea surface. Spatial gaps in coverage (array divergence) are also minimized by lowering surface time. Deep Argo floats can now reach depths of 6000 m and biogeochemical Argo floats carry additional sensors for dissolved oxygen, nitrate, pH, and bio-optical parameters. Argo's CTDs are more accurate and stable than imagined in 1998, thanks to a strong partnership of Argo scientists, engineering teams, and commercial suppliers. These and other innovations have originated in many different national programs and have been freely described and shared via the AST.

The large size of each US Argo institutional contribution, and of several other national programs, enabled experimentation with different approaches to Argo float production and technology improvement. Enough floats are produced by each of these partners to demonstrate float robustness, to detect design and production faults, and to provide confidence in float performance. At the same time, the numbers are sufficient to realize economies of scale in production and deployment. Distinct institutional approaches range from design of new float models and full float fabrication, to partial assembly and thorough testing of commercial floats, to simple testing of warrantied commercial floats. Three different commercial US float vendors are now partners with US Argo float providers, and other commercial float models developed in France, Japan, Germany, and Canada have been or are being deployed by national Argo Programs. This diversity of approaches has resulted in the advances described above and others, making the Argo Program more cost-effective as fewer new floats are needed to sustain the array and less frequent deployments are required for re-seeding.

The first Argo floats were deployed in 1999 and the early years focused on regional pilot arrays. By the end of 2002, there were over 500 operational Argo floats. Pilot arrays were expanding and by late 2004 Argo deployments were being made globally, with 1500 operational floats. In late 2007 the milestone of 3000 operational floats was achieved. The rate of increase slowed as floats aged and attrition began to balance new deployments. The number of operational floats reached 4000 in 2016 and has remained near that level (Fig. 4.2). Because floats have a relatively long lifetime the array is not immediately sensitive to transient interruptions in deployment such as caused by the COVID-19 disruption of RV traffic.

The spin-up phase of the Argo array in 1999–2007 was coordinated by the AST together with the partnering national agencies. All Argo Programs had national regional priorities for their deployments, and these had to be integrated in a way that would create uniform global coverage. For extending coverage to especially remote regions in the southern hemisphere, having little or no RV traffic, a regional partnership of US, New Zealand, and Australian Argo Programs provides the necessary floats and deploys them from a small N.Z. RV. Without these targeted deployments, totaling over 1900 floats by RV Kaharoa beginning in 2004, global coverage could not have been achieved.

International governance: Argo and JCOMM

During Argo's spinup phase, W.S. Wilson served as US Representative to the Intergovernmental Oceanographic Commission (IOC) of UNESCO, an organization facilitating international oceanographic collaboration.

Wilson's advocacy enabled the IOC to become an international forum promoting Argo to multiple countries not directly involved in the program. For example, in the Pacific where much of the ocean area is comprised of Exclusive Economic Zones (EEZs), the IOC helped facilitate an informal agreement in 2002 with the South Pacific Applied Geoscience Commission (SOPAC), enabling floats to be launched in the EEZs of its 13 member island nations, a sizable fraction of the collective EEZ of the Pacific. The SOPAC "concurrence" confirmed acceptance by the member nations, primarily through their meteorological and fisheries agency heads, of deployment of Argo floats in the EEZs.

A key element for Argo's international acceptance is the transparency of Argo operations and planning, including float deployments and drift. The World Meteorological Organization (WMO), acting through the Joint Technical Commission for Oceanography and Marine Meteorology (JCOMM[6]), tracks the position of all Argo floats and makes the location data publicly available. The float tracking system, including future deployments, encourages international support for Argo's global coverage and ensures Argo's consistency with the U.N. Convention on the Law of the Sea. About 30% of the global ocean is inside EEZs or in other maritime zones such as the Antarctic Treaty Zone. For Argo to sample the global ocean, it is critical to have access for float deployment, drift, and data collection inside these regulated areas.

The IOC, at its General Assembly in 1999, adopted Resolution XX-6, accepting Argo as "an important contribution to the operational ocean observing system of GOOS as well as a major contribution to CLIVAR and other scientific research programmes". That same year the WMO, at its 13th Congress, endorsed Argo in the same context. In 2008, the IOC Executive Council adopted Resolution EC-XLI.4, including a set of guidelines "regarding the deployment of profiling floats *in the high seas* within the framework of the Argo Program." Further, the IOC recognized the long-term need for Argo: "A full array of more than 3000 Argo profiling floats has been deployed in the world ocean and the Argo Project is now working and should be sustained in the future as the Argo Program." In 2018 the IOC Executive Council approved the addition of 6 biogeochemical parameters to Argo floats, subject to the same guidelines defined in EC-XLI.4.

---

[6] The JCOMM Observations Program Support (JCOMMOPS) center was recently renamed OceanOPS.

The high-level recognition by the IOC helped to raise the international visibility of Argo and the acceptance of Argo floats drifting throughout the global ocean. However, the deployment of Argo floats inside EEZs remains a complicated topic even today, with differing views over Argo's predominant role as either operational oceanography or marine science research. Consequently, float deployments in some EEZs are less than Argo targets.

The Argo data management system

A critical element of Argo's success and sustainability is its data management and distribution system, which occupies central roles in the Argo partnership. Argo's innovative data system, as with its profiling float technology, evolved from WOCE precursors. WOCE maintained Data Assembly Centers (DACs) for the WOCE Hydrographic Project, Upper Ocean Thermal data (mostly XBT profiles), profiling floats, and other data types. A shortcoming of historical datasets is sparse metadata and technical data, which are needed for accurate interpretation. The WOCE DACs took steps toward correcting this problem. Argo subsequently developed a new data management system, taking lessons from the WOCE DACs, and including metadata, technical data, and quality flags along with the measurements of temperature, salinity, pressure, and trajectory. Central in this initial system design were R. Keeley (Canada) and S. Pouliquen (France).

An Argo Data Management Team (ADMT) was formed as a subcommittee of the AST, with the first ADMT meeting held in 2000 at IFREMER/Brest. As with the float hardware, many nations, through their funded national Argo Programs, have contributed resources and expertise to building and improving the Argo data system. The AST had urged all national Argo Programs to include in their proposals 15% dedicated to data management. Most programs followed this guideline. All Argo data are available in the self-documenting NetCDF format, from either of the two Argo Global Data Assembly Centers that synchronize their data holdings to ensure continuous access. The Argo data system has been a template for data systems developed for other elements of the observing system.

A cornerstone of the Argo data system is free and open data availability. This was recognized by the AST as essential for maximizing the value of Argo data. Moreover, increased support for Argo by all nations is achieved by transparency and sharing. Anyone in the world with an internet connection has the same free and open access to Argo data as an Argo participant. The example set by Argo and other programs has become accepted practice in sustained ocean observations. The success of Argo's open data policy is illustrated by Argo data having been used in over 5000 research papers.

The breadth of Argo's user community made it essential to release both a near-real-time (RT) dataset and a research-quality delayed mode (DM) dataset. RT data, with over 90% of profiles available within 12 h of collection, are required by operational centers for ocean and coupled model initialization and data assimilation. DM data are subject to expert examination to ensure the highest quality for research applications. Argo provides documentation[7] of its RT and DM procedures and holds technical workshops to ensure that the DM dataset is uniform and consistent across the many national Argo Programs. Moreover, Argo data must be consistent through Argo's temporal domain, and consistent with standards set by Argo CTD calibrations and by highest quality GOOS temperature and salinity datasets, such as shipboard repeat hydrography.

The sustainability of Argo

The sustainability of Argo depends on the ability of the international partnership to provide and deploy a sufficient number and spatial distribution of Argo floats to maintain the global array. Since 2004, annual Argo deployments have exceeded 800 floats in most years. Deployment locations are prioritized by the national programs, but with consideration to filling gaps in global coverage. The large number of international partners has tended to smooth out the year-to-year variations in funding that occur in individual national programs.

The cost of Argo is, of course, an important element in sustainability. Over the lifetime of a single float the total expenditure is roughly US$35,000, including the float cost, with CTD, and all operating costs (communications, global logistics, deployment, data management, engineering). So, for international Core Argo to provide 800 floats per year the cost is about US$28,000,000. This is about US$200 for each of the ~140,000 temperature/salinity profiles per year. Inflationary increases in cost have been largely offset by the technology improvements mentioned above, including increases in float lifetime.

[7] http://www.argodatamgt.org/Documentation.

The continuing support of Argo by many partner nations reflects the high value of the Argo dataset in research, climate assessment and forecasting, education, and operational oceanography. By 2010, over 200 research publications per year were using Argo data, increasing to over 400 per year since 2016[8]. The breadth of topics of Argo research continues to grow, including ocean heat content and warming, water mass formation and spreading, air-sea exchange, ocean circulation, mesoscale variability, mixing, and more. Assessments of the state of the climate system, by the Intergovernmental Panel on Climate Change, the Bulletin of the American Meteorological Society, and others, rely on Argo data for documenting the changing ocean state. As a climate dataset, Argo's value grows with the time span of the observations. Argo is the primary dataset used for assessment of ocean heat content and for global and regional warming of the Earth's climate system. A lengthening dataset also enhances Argo's value in education and operational modeling. Argo data have been used in over 300 PhD theses, as well as in high school, undergraduate, and graduate classrooms. Even some primary school curricula and enrichment programs incorporate Argo in learning about oceans, weather, and climate[9]. All operational ocean and coupled models are using Argo data. Forecast and ocean reanalysis models depend on Argo for initialization or data assimilation.

The continuing participation and renewal of key personnel in Argo is another hallmark of sustainability. Scientists analyzing Argo data and graduate students using it in their PhD theses are sources of new expertise for Argo. Engineering teams include both long-term Argo participants with sharp memories for what works and what doesn't and recent graduates carrying new ideas for tackling old problems. Data managers continue to be among Argo's most valuable and creative contributors.

Another element of the partnership, and another source of sustained support for Argo, is the academic institutions and government laboratories that house Argo teams. This dates to the beginnings of Argo when the program could never have begun without institutional salary support for scientists pursuing unconventional ideas. Institutional support has been steady throughout Argo's history, providing support for academics, and space and facilities for the laboratories that are Argo's heart. Moreover, institutional roles go beyond material and personnel support. The institutions play critical

---

[8] https://argo.ucsd.edu/outreach/publications/bibliography/.
[9] https://argo.ucsd.edu/outreach/education-materials/.

roles in molding national resources and agency priorities. Without the strong pull of agencies for Argo implementation, reinforced by institutional voices, Argo could not have achieved the size needed for a global array.

The Argo Program continues to evolve even as the original Core Argo array is sustained into its 3rd decade. Major new elements include Deep Argo, extending coverage beyond 2000 m to the ocean bottom (Johnson et al., 2015; Roemmich et al., 2019a) and increasing CTD accuracies to resolve deep-ocean variability. Biogeochemical (BGC) Argo, is adding new sensors for dissolved oxygen, pH, nitrate, and bio-optical properties (Claustre et al., 2020). Implementation of Deep Argo will close the water column components of the ocean's heat, freshwater, and sea level budgets. BGC Argo will track ocean acidification, deoxygenation, the carbon budget, and other biological and geochemical signals on regional and global scales. In addition to these new elements, enhanced global Argo coverage is underway. This includes doubling of float density in the tropical Pacific ($10°S-10°N$) and in energetic western boundary regions, together with coverage enhancements in marginal seas and high latitude oceans with seasonal ice cover. The "OneArgo" design (previously "Argo 2020", Roemmich et al., 2019b) would require a total array size of 4600 floats, including 2350 Core Argo, 1250 Deep Argo, and 1000 BGC Argo floats.

There is strong community support for both Deep and BGC Argo expansions. Argo's international partnership and infrastructure are also expanding to include these valuable new elements, with each of these programs requiring new resources comparable to Core Argo itself. What makes this challenging is not only the support level but also that the increases must be sustained far into the future, as for Core Argo. Even as regional pilot projects, Deep and BGC Argo are producing fundamental new insights into ocean circulation (Zilberman et al., 2020; Le Traon et al., 2020), intra-seasonal to decadal variability, and the Earth's climate system (Johnson et al., 2019).

Conclusion

This case study is focused on the Argo partnership and its role in the program's success and endurance, but with the caveat that partnership alone was not sufficient. The US National Research Council (2015) characterized the reasons for Argo's success. To paraphrase, Argo succeeded because it crystallized a shared concept of global ocean observing with a technology capable of carrying the concept to a practical stage.

Meanwhile, continuing evolution, transparency, and open exchange of data were sources of renewal. Argo's partnership has been critical, built as it is on a solid foundation of science and technology, with clear and measurable common goals.

The Argo Program was created from the bottom up. It began by asking "what if we could deploy a global array", and then began attracting programmatic and agency attention for that objective. Argo has not been subjected to outside direction, and having a degree of autonomy has given Argo the freedom to be steered by scientific objectives and technology progress. Its unfettered development means that Argo partnerships are informal, built on shared objectives among the partners rather than by formal commitments. Mutual trust and understanding are hallmarks of the program and important factors in its success. As Argo matures it is increasingly impacted by the landscape of GOOS. Argo is not a complete observing system. It has key synergies with many other GOOS satellite and *in situ* elements that must be maximized.

Of course, Argo has vulnerabilities as well as strengths. The inability to achieve a global consensus on guidelines for float deployment inside EEZs has reduced coverage in a few EEZs and is inefficient. For deployment opportunities, Argo is largely dependent on the RV fleets, and these vessels do not provide sufficient traffic in the southern hemisphere and the Arctic to sustain Argo. Chartered vessels have mitigated but not yet solved this problem. Argo's single dominant source of CTDs is a vulnerability that has contributed to gaps in supply and concerns about sensor stability and accuracy. The AST is addressing this last issue by defining a procedure for evaluation and adoption of new sensors, to enable new CTDs to be introduced without jeopardizing the robustness of data from the array.

Perhaps Argo's greatest present vulnerability is also its greatest present strength, and that is the major expansion into Deep Argo and BGC Argo. Rapid growth inevitably brings challenges to sustaining the partnership and the core principles of the original Argo Program, as many new scientific partners and technology experts are entrained. But if Argo can overcome the challenges of expansion, its value will grow enormously as it targets the full depth of the global ocean and the multi-disciplinary aspects of the oceanic roles in Earth's climate system.

A new chapter in the Argo partnership story will be its role in the United Nations Decade of Ocean Science for Sustainable Development. By providing critical global ocean observations, Argo sits at the base of the value chain for many of the U.N. Decade's Sustainable Development

Goals, including Quality Education, Innovation and Infrastructure, Climate Action, and Life Below Water. Argo will first help to define and then to realize the U.N. Decade's mission of "the science we need for the ocean we want".

Acknowledgments

Argo data are collected and made freely available by the International Argo Program and the Argo National Programs that contribute to it (http://www.argo.ucsd.edu, http://ocean-ops.org, http://doi.org/10.17882/42182). The authors gratefully acknowledge support from their respective Argo National Programs or national agencies, who have made the Argo Partnership possible. Though many of these contributions are noted in the text of this Chapter, we also recognize many others whose names do not appear in publications but whose contributions on land and sea helped bring the vision of Argo to fruition.

References

Argo Science Team, 1998. On the Design and Implementation of Argo: A Global Array of Profiling Floats. International CLIVAR Project Office Report 21, GODAE Report 5. GODAE International Project Office, Melbourne, Australia, p. 32.

Boyer, T.P., Antonov, J.I., Baranova, O.K., Coleman, C., Garcia, H.E., Grodsky, A., Johnson, D.R., Locarnini, R.A., Mishonov, A.V., O'Brien, T.D., Paver, C.R., Reagan, J.R., Seidov, D., Smolyar, I.V., Zweng, M.M., 2013. World Ocean Database 2013. National Oceanographic Data Center, Silver Spring, MD, p. 208 (NOAA Atlas NESDIS, 72).

Claustre, H., Johnson, K.S., Takeshita, Y., 2020. Observing the global ocean with biogeochemical-Argo. Annu. Rev. Mar. Sci. 12, 23–48.

Davis, R.E., Sherman, J.T., Dufour, J., 2001. Profiling ALACEs and other advances in autonomous subsurface floats. J. Atmos. Ocean. Technol. 18 (6), 982–993.

Gould, W.J., 2005. From swallow floats to Argo—the development of neutrally buoyant floats. Deep Sea Res. Part II Top. Stud. Oceanogr. 52 (3–4), 529–543.

Gould, W.J., 2003. WOCE and TOGA-the foundations of the Global Ocean Observing System. Oceanography, Washington DC. Oceanogr. Soc. 16 (4), 24–30.

Johnson, G.C., Purkey, S.G., Zilberman, N.V., Roemmich, D., 2019. Deep Argo quantifies bottom water warming rates in the Southwest Pacific Basin. Geophys. Res. Lett. 46 (5), 2662–2669.

Johnson, G.C., Lyman, J.M., Purkey, S.G., 2015. Informing deep Argo array design using Argo and full-depth hydrographic section data. J. Atmos. Ocean. Technol. 32 (11), 2187–2198.

Le Traon, P.Y., Rienecker, M., Smith, N., Bahurel, P., Bell, M., Hurlburt, H., Dandin, P., 2001. Operational oceanography and prediction − a GODAE perspective. In: Koblinsky, C.J., Smith, N.R. (Eds.), Observing the Oceans inthe 21st Century.

Le Traon, P.Y., 2013. From satellite altimetry to Argo and operational oceanography: three revolutions in oceanography. Ocean Sci. 9 (5), 901–915.

Le Traon, P.Y., D'Ortenzio, F., Babin, M., Leymarie, E., Marec, C., Pouliquen, S., Thierry, V., Cabanes, C., Claustre, H., Desbruyères, D., Lacour, L., 2020. Preparing the new phase of Argo: scientific achievements of the NAOS project. Front. Mar. Sci. 7, 838.

McPhaden, M.J., Busalacchi, A.J., Cheney, R., Donguy, J.R., Gage, K.S., Halpern, D., Ji, M., Julian, P., Meyers, G., Mitchum, G.T., Niiler, P.P., 1998. The Tropical Ocean-Global Atmosphere observing system: a decade of progress. J. Geophys. Res.: Oceans 103 (C7), 14169—14240.

National Research Council, 2015. Sea Change: 2015-2025 Decadal Survey of Ocean Sciences. The National Academies Press, Washington, DC.

Rhein, M., Rintoul, S.R., Aoki, S., Campos, E., Chambers, D., Feely, R.A., Gulev, S., Johnson, G.C., Josey, S.A., Kostianoy, A., Mauritzen, C., 2013. Observations: ocean. In: Climate Change 2013: The Physical Science Basis. Contribution of Working Group I to the Fifth Assessment Report of the Intergovernmental Panel on Climate Change. Cambridge University Press, Cambridge, United Kingdom and New York, NY, USA.

Roemmich, D., Sherman, J.T., Davis, R.E., Grindley, K., McClune, M., Parker, C.J., Black, D.N., Zilberman, N., Purkey, S.G., Sutton, P.J., Gilson, J., 2019a. Deep SOLO: a full-depth profiling float for the Argo Program. J. Atmos. Ocean. Technol. 36 (10), 1967—1981.

Roemmich, D., Alford, M.H., Claustre, H., Johnson, K., King, B., Moum, J., Oke, P., Owens, W.B., Pouliquen, S., Purkey, S., Scanderbeg, M., et al., 2019b. On the future of Argo: a global, full-depth, multi-disciplinary array. Front. Mar. Sci. 6, 439.

Siedler, G., Church, J., Gould, J. (Eds.), 2001. Ocean Circulation and Climate: Observing and Modelling the Global Ocean (International Geophysics Series, vol. 77. Academic Press, San Fransisco CA, USA, p. 736.

Smith, N.R., Lefebvre, M., 1997. Monitoring the oceans in the 2000s: an integrated approach. In: International Symposium: "the Global Ocean Data Assimilation Experiment". GODAE) Biarritz, October, pp. 15—17.

Stammer, D., Wunsch, C., Giering, R., Eckert, C., Heimbach, P., Marotzke, J., Adcroft, A., Hill, C.N., Marshall, J., 2003. Volume, heat, and freshwater transports of the global ocean circulation 1993—2000, estimated from a general circulation model constrained by World Ocean Circulation Experiment (WOCE) data. J. Geophys. Res.: Oceans 108 (C1), 7-1.

Zilberman, N.V., Roemmich, D.H., Gilson, J., 2020. Deep-ocean circulation in the southwest Pacific Ocean interior: estimates of the mean flow and variability using Deep Argo data. Geophys. Res. Lett. 47 (13) e2020GL088342.

CHAPTER 5

The Marine Arctic Ecosystem Study partnership: planning, implementation and lessons learned

Guillermo Auad[a] and Francis K. Wiese[b]

[a]Office of Policy and Analysis, Bureau of Safety and Environmental Enforcement, U.S. Department of the Interior, Sterling, VA, United States; [b]Stantec Consulting Services, Inc., Anchorage, AK, United States

Introduction

A fresh set of eyes focusing on the marine ecosystems of the Alaskan Arctic led to the identification of information needs for the United States (US) Bureau of Ocean Energy Management (BOEM) to inform its decisions on offshore energy in the Beaufort Sea. Subsequently, common and complementary information needs were identified by a group of scientists from different US Federal agencies during 2011 and 2012. Arctic research policy promoting interagency and international collaborations added further momentum to this effort (US Arctic Research Commission, 2013). In this chapter, we address the planning, implementation, and lessons learned throughout the 2011–20 timeframe of the Marine Arctic Ecosystem Study (MARES).

Project planning

Led by BOEM, and facilitated by the National Oceanographic Partnership Program (NOPP), nearly a dozen planning and coordination meetings, some with over 30 attendees, took place in 2012. At these meetings, it was possible to communicate, among funding partners, who needed what and how much each partner could contribute and how. NOPP, created by an act of the US Congress in 1997, was the perfect forum to help facilitate this process as its core mission is to facilitate partnerships between federal agencies, academia, and industry to advance ocean science research and education. In general, NOPP projects span a broad range of topics in oceanographic research, including environmental monitoring, ocean exploration, earth systems modeling, technology development, and marine resource management. Through NOPP, federal agencies can leverage resources to invest in priorities that fall between agency missions or that are too large for any single agency to support.

These year-long conversations addressed science, resources, geographical area, timing, and management aspects, including data management. From the literature review conducted by the lead funding partner, an initial list of objectives was drafted, additional objectives were provided by all involved in the planning meetings, and a single integrated list of objectives was finalized from which other discussions emerged about management and technology; the *Marine Arctic Ecosystem Study*, or MARES, was born.

Supported by the call for integration and interagency coordination emphasized by the Arctic Research Policy Act of 1984, the planning and coordination group considered several strategies to integrate knowledge across disciplines, e.g., biophysical, and domains, e.g., ocean-atmosphere and the steps that would be needed to get from concept to implementation (see Fig. 5.1). It was agreed that integration would be facilitated by the type of procurement selected and the number of planned solicitations and awards.

Different approaches from past large studies were considered, ranging from one solicitation per agency and multiple awards, to one large combined solicitation; criteria including speed of implementation and study integration were discussed. The multiple solicitation approach was considered faster during implementation but fragmented in terms of outcomes without much integration across component projects with potentially different timelines. In the end, it was agreed that the best way forward to implement MARES would be one comprehensive and consistent

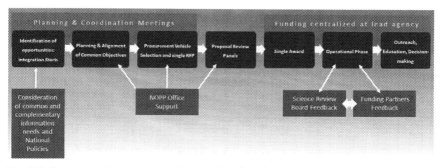

Fig. 5.1 Summary of the steps taken to develop the MARES solicitation and subsequent implemented and optional tasks and actions.

solicitation with one award to an integrated inter-disciplinary team. It was also agreed that NOPP would provide the needed coordination to prepare a large solicitation document as well as to host and finance one of the three review panels needed to evaluate proposals. The other two panels included a Federal-only ad-hoc review group and an executive panel chaired by a representative from the lead funding agency. This latter panel included representatives from five federal agencies (Bureau of Ocean Energy Management, Office of Naval Research, National Science Foundation, Marine Mammal Commission, and National Oceanographic and Atmospheric Administration) plus observers.

To arrive at a mutually agreeable and sufficiently detailed statement of work and subsequent request for proposals (RFP) by all funding partners, a series of all hands and one-on-one meetings were held between the lead agency (BOEM) and each of the other potential funding partners during 2012 and early 2013. Then the lead agency drafted a request of proposals document with input from scientists, managers, and contracting officers in different programs and offices. A series of reviews followed by all funding partners. This led to one single RFP agreed to by all.

Because the lead agency had no grant authority, the conversation then focused on selecting a government contract or a cooperative agreement as the procurement vehicle. Once more, the goal of attempting to deliver well-integrated scientific results was the driver which pointed to a modular approach to integrating in-situ observations of biological, chemical, and physical variables in the ocean, atmosphere, and land, with those gathered from numerical models and satellite observations. An Indefinite Delivery Indefinite Quantity (IDIQ) government contract was selected because it allows individual components or "task orders" to be aligned under an overarching contractual architecture. These task orders provide great flexibility in that each of them can be (a) funded by a subset of funding partners, (b) conducted by a sub-set (or all) of the research partners within the selected team, and (c) be integrated at the back end by the funded inter-disciplinary team (Fig. 5.2).

Project implementation

An integrated solicitation was issued to the public on June 13, 2014, three years after the idea was first conceived. Partner organizations initially included the United States Arctic Research Commission (USARC), the National Oceanic and Atmospheric Administration (NOAA), the Office of

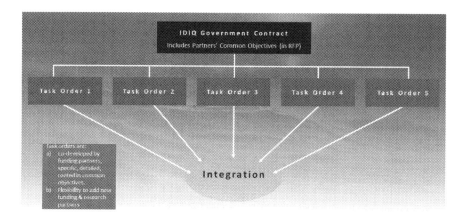

Fig. 5.2 Structure of the main binder in MARES: The IDIQ government contract structure. It is a modular approach where each Task Order is an independent but integrable element of the project. Funding for all task orders was concentrated in the lead agency even though it did not contribute fully or totally to fund some of these components. Every component activates different funding and research partners. Each partner can participate in one or more components. The last task order in MARES served as the integrative final step.

Naval Research (ONR), the United States Integrated Ocean Observing System (US IOOS), the National Science Foundation (NSF), the US Coast Guard (USCG), the United States Geological Survey (USGS), the Marine Mammal Commission (MMC), and Shell Oil Company. With an anticipated funding ceiling of $25 million US dollars over 5 years and multiple Task Orders, the stated goal for the research was to advance the overall knowledge of the structure and function of the Arctic marine environment in the Beaufort Sea, including identification of its main controlling drivers and agents of change, aligning with the recommendations of the 2011 USGS Arctic report, the National Ocean Policy guidelines, and IARPC 5-year plan, the latter a mandated US Arctic research policy. The RFP called for one integrated team to conduct retrospective analyses, field studies, modeling, and community-based monitoring along the continental shelf in the Beaufort Sea, spanning atmospheric sciences, physical, chemical, and biological oceanography, marine mammal research, incorporation of local and traditional knowledge, and data management.

After the aforementioned 3-layer reviews of the proposals received, Stantec Consulting Services Inc. (Stantec) was selected as the lead research partner in September 2014. Stantec's internal team consisted of marine scientists and modelers in the US and Canada, including team members from several academic, private, state, and tribal organizations in the US and Canada. Forming this research partnership was very much a social process based on internal Stantec expertize and networks, and existing relationships between the Principal Investigator and other scientists and community members across Canada and the US. The team was structured to address the different topics, disciplines, and approaches detailed in the RFP and originally included 65 people from 21 institutions. Once the contract was awarded, all participants were notified and put on the subcontractor list. As Task Orders were issued over a period of 5 years, detailed proposals and cost estimates had to be developed for each Task. Once awarded, subcontracts were set up between Stantec and each partner institution included in each of the Tasks. In addition, in-kind and other partnerships were built on trust and previous relationships to maximize the opportunity and find synergies with logistics and research endeavors with other programs in the region, e.g. the annual Canadian Coast Guard cruise, the ArcticNet program, and seal tagging by the North Slope Borough Department of Wildlife Management.

Throughout the project life, five task orders were issued to address project and data management, tagging of marine mammals with novel sensors, use of autonomous vehicles, collection of water and sediment samples to characterize carbon cycling processes in the Eastern Beaufort Sea, and deployment of several ecosystem moorings located at different depths and heavily instrumented from top to bottom (Fig. 5.3).

Through these task orders, 16 institutions were brought together to conduct the research needed by the funding partners (Table 5.1).

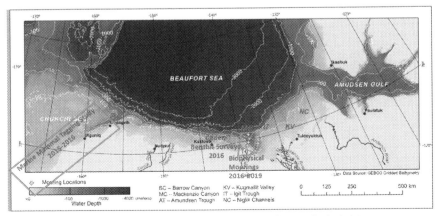

Fig. 5.3 MARES study locations in the Beaufort and Chukchi seas.

Table 5.1 MARES partners, sectors and roles.

Partner	Contribution	Sector and country
Bureau of Ocean Energy Management (lead)	Funding	Federal government, US
Office of Naval Research	Funding	Federal government, US
US Arctic Research Commission	Funding	Federal government, US
Dutch Shell Oil	Funding	Private sector, US
US Coast Guard	In-kind resources	Federal government, US
National Oceanographic and Atmospheric Administration	In-kind resources	Federal government, US
Smithsonian Institution	In-kind resources	Federal government, US
Arctic Net	In-kind resources	NGO, Canada
Stantec Consulting Services (lead)	Research and management	Private sector, US and Canada
Kavik-Stantec	Consultations and permitting	Aboriginal, Canada
Woods Hole Oceanographic Institution	Research	NGO/Academic, US
Virginia Institute of Marine Science	Research	Academic, US
University of Alaska, Fairbanks	Research	Academic, US
University of Washington	Research	Academic, US
Old Dominion University	Research	Academic, US
ASL Environmental	Research	Private sector, Canada
Seastar Biotech	Research	Private sector, Canada
Monterey Bay Aquarium Institute	Research	NGO, US
North Slope Borough	Research, traditional knowledge	Tribal, US
DFO Fisheries and Oceans Canada	Research and logistics	Federal government, Canada
Alaska Ocean Observing System	Data management	NGO, US
AXIOM services	Data management	Private sector, US
Alaska Department of Fish and Game	Permit	State government, US
Royal Canadian Coast Guard	Logistics	Federal government, Canada

Note that the research partners listed are only those who were funded, several more were part of the original proposal team when the envisioned scope was larger than what was ultimately implemented.

All funding from Federal agencies and private sector partners, excluding in-kind resources, was concentrated in the lead funding agency (BOEM), which issued different task orders as funding became available from the different partners. For example, the second task order concentrated funds from the Office of Naval Research and Dutch Shell Oil. Even though this component had no funding from BOEM, it managed those funds and awarded this task order from the single focal point defined by the IDIQ contract. Other task orders had funding from BOEM and the US Arctic Research Commission, while others included resources from BOEM, Dutch Shell Oil, and the US Coast Guard. Some Task Orders were administered by different Contracting Officer Representatives (CORs) but were all kept under one overall COR and Contracting Officer, maintaining continuity throughout the program.

Different "binders" were used to formalize linkages among the different research and funding partners as elaborated on above. Among the funding partners, interagency agreements were used for most agency-to-agency collaborations, while a letter of commitment was used between the lead agency and Dutch Shell Oil.

Although the support from some potential funding partners eventually fell through (more on this below) leading to a reduction in scope, i.e. no resources were available to support all project components as originally envisioned, including atmospheric sciences, a comprehensive research cruise, community-based monitoring, and modeling, over 30 variables were measured in the eastern Beaufort Sea (Table 5.2, Fig. 5.3) to provide a more holistic view than previously undertaken in this area and vicinity. Although the overall MARES budget was around ten million US dollars, all sources considered, the reduction in scope can be traced to reasons which were independent of any budget limitations. Among these were personality issues, management decisions, as well as miscommunications.

Lessons learned

Sustainability

MARES was scientifically and programmatically successful and resilient to challenges over its 6-year duration due to the presence of key underpinnings enabling resiliency in the partnership system: connectivities, flexibility, diversity, and redundancy. Specifically this was achieved through:
- A lead agency with a strong internal champion able to build bridges and partnerships across federal agencies toward a common goal (connectivity).

Table 5.2 Over 30 physical, chemical, and biological variables were observed and used in MARES.

Variable	Platform/s
Temperature	Moorings, glider, satellite, animal-mounted sensors, CTD casts
Salinity	Moorings, glider, animal-mounted sensors, CTD casts
Zooplankton abundance	Moorings, glider
Fish abundance	Moorings
Marine mammal vocalizations	Moorings
Dissolved carbon dioxide	Moorings, water samples
Oxygen	Moorings, glider
Nitrates	Moorings, water samples
Ammonia	Water samples
Silicate	Water samples
Phosphate	Water samples
Fluorescence (for Chlorophyll)	Moorings, glider, water samples, animal-mounted sensors
Photosynthetically active radiation (PAR)	Glider, mooring
Carbon dissolved organic matter	Glider
Turbidity	Glider, moorings, satellite
River discharge volume	Gauge data
Meridional velocity (north-south)	Moorings, ocean model
Zonal velocity (east-west)	Moorings, ocean model
Vertical velocity	Moorings, ocean model
Seal positioning and 3D movement	Animal-mounted sensors
Fatty acids	Sediment samples
Carbon isotopes	Sediment samples
Dissolved inorganic carbon	Sediment samples
Sterol biomarkers	Sediments samples
Benthic meiofauna	Sediment samples
Sediment grain size	Sediment samples
Sea ice coverage	Mooring, satellite
Sea ice thickness	Moorings
Sea ice movement	Mooring, satellite
Air temperature	Met Station
Atmospheric pressure	Met Station
Wind speed and direction	Met station and atmospheric circulation model (ERAS, NCEP)

- A lead research partner able to build and maintain an integrated interdisciplinary multi-institutional team and contractually, financially, and scientifically implement a highly complex project in a harsh and remote environment (connectivity, flexibility).
- Formalized agreements and personal relationships within and between funding and research partners (connectivity, flexibility, and redundancy).
- A modular (Task Order) but integrated structure (Fig. 5.2) within one solicitation and one award approach (diversity, connectivity, flexibility).
- Continuous and fluid communication at said focal point between the lead funding and research partners, including open channels of communication with other funding and research partners based on trust and personal relationships (connectivity, flexibility, diversity).

In addition to these key elements of resilience, the addition of several monetary, in-kind, and personal resources were instrumental in sustaining the MARES partnership for over 6 years. One such resource was the passionate dedication to scientific research of everyone involved, and this was a fundamental driver providing vital fuel to the overall enterprise. We next analyze each of these elements of resilience in more detail.

Lead agency and lead research partner as champions

The single focal point of the MARES partnership was shared by BOEM, the lead funding agency, and Stantec, the lead research partner. This focal point shielded Stantec from receiving funds and directives from multiple sources at different points in time and favored the integrative process sought from the beginning in the planning stages. Away from this focal point, research and funding partners were linked together similarly to how roots and branches are connected to the trunk of a tree. The MARES partnership can then be represented or visualized through a tree-like conceptual model. Fig. 5.4 illustrates this where the funding partners (roots) acquire funding (nutrients) within the funding environment (soil) and funnel those through the single focal point (trunk) to the rest of the tree (partnership).

The research partners (branches, leaves) acquire information (air, oxygen) from the research environment (atmosphere) to be used to maintain the life of the entire tree and often to propel growth over time. By using this conceptual model one can justify the decision of using one solicitation

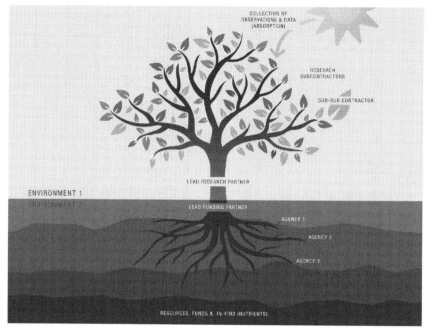

Fig. 5.4 Conceptualization of the MARES partnership model distinguishing between environments (funding and research) and emphasizing the importance of a single focal point (trunk) and the flow of resources and data and information (nutrients, gases) in opposite directions, as in a tree.

with one award over other available choices. This design, a tree, has a structure that transports different properties in opposite directions (nutrients from the roots and gases from the leaves), and it has been used by nature many times over, e.g. lungs, deltas.

Formalized agreements

The different types of "bindings" were a key component to make this project sustainable and resilient to all the challenges encountered, and in some cases, the binding in question was signature-less, i.e., trust. Once the MARES contract to Stantec was awarded, and all the subcontracts to the research partners were in place, all these agreements (including signature-less ones), contracts and subcontracts kept the partnership together for 6 years through flexible binders. These, allowed to activate different researchers on an as needed basis and allowed adjustments to pre-agreed dates or research platforms due to, e.g., harsh weather conditions.

For example, one of these resulted in changing the location of the study area which in turn attracted the interest of Canadian scientists and their respective organizations, further increasing the MARES bi-national investments in marine science and the size of the overall research team.

Modular approach

The modular (Task Order) approach was arguably one of the biggest successes of MARES as it had not been attempted before at this scale and in the Arctic; it provided flexibility to funding and research partners and was a key factor in facilitating integration. The IDIQ type of contract functioned as an integrative framework that facilitated the activation of specific funding and research partners for each component (or task order) of MARES. The modular approach, the branches of the tree, allowed managers to isolate those issues in one module rather than affecting the entire project. Addressing these situations within one module of the project, and having a single focal point for management and coordination, streamlined and facilitated corrective and mitigating actions to continue with the MARES research agenda. The tree-like architecture was not only effective for addressing these challenges but also for taking advantage of opportunities, such as the use of a Canadian icebreaker when it became more difficult and expensive to use US-flagged vessels. This flexibility was also useful when very unusual sea ice conditions in the eastern Beaufort Sea prevented the recovery of one mooring in the Fall of 2018. A contingency plan was quickly implemented and the mooring in question was recovered a year later with the bonus of having some time series extending 3, 6, and even 11 months beyond the planned observational period.

At the same time, this modular approach was not without challenges. From a research team perspective, this meant that not all partners were active in all task orders, requiring extra effort to keep people interested, informed, and involved for when they were needed again in later tasks and the overall project integration. On the program and financial management side, this approach also meant that each task order had to be administered separately, including not being able to spill over money across task order contracts. The extra administrative burden was ultimately offset by the benefits outlined above. Professionalism, trust, personal relationships, and the passion for the pursuit of science and excellence in project delivery were all key to achieving the needed continuity across task orders.

Communication

The issue of communication, a form of connectivity enhancing resilience, was crucial in MARES, both before and after award, within the funding agencies, within the research team and other partners, and between the funding agency and the research team. This communication included about 12 pre-awarded coordination and planning meetings by the funding agencies and Shell, and weekly communications among the research team led by Stantec after the award. Numerous planned and unplanned meetings were held between the lead funding and research organizations. To reduce the risk of miscommunications, the single focal point at the funding-research intersection only had two people (one from BOEM, one from Stantec) coordinating and adjusting key managerial/funding aspects, and two other people (one from BOEM, one from Stantec) coordinating technical/scientific aspects. This approach reduced the risk of miscommunications and saved time and, because of the reduced number of actors, it also quickly built trust and momentum to address future challenges together.

Similarly, the science team, as well as the Stantec science leads, program, and financial managers met monthly and bi-weekly throughout the program to stay integrated, informed, and on top of the programmatic requirements.

Flexibility and connectivity (in quantity and type), as well as values such as trust, respect, and solidarity, were all essential from day one. Nearly one hundred very diverse people were involved in MARES at different stages and in different capacities and roles. This collaborative social process is the essence of science (Merton, 1973), and led to the overcoming of many obstacles and to the convergence of all activities on many findings, some of which were firsts. Remarkably, one hundred percent of all instruments and sensors were recovered and over 95% of data recovery was achieved.

Trust and continuity

Despite all the front-end meetings and coordination, some funding partnerships and associated resources fell through. The reasons for this fall through are many. Some potential funding partners changed their representatives before the MARES coordination and planning group and had poor internal communication, making assumptions rather than asking; others were attending the meetings to see if they could fund their government scientists rather than providing resources. From the planning and coordination perspective, more specific one-on-one meetings between the lead agency and potential funding partners would have likely paved the way to procure additional resources.

At the 2020 Ocean Sciences Meeting in San Diego, California, MARES received the 2019 Excellence in Partnering Award from the NOPP (https://www.nopp.org/2020/2019-excellence-in-partnering-award/). It is therefore relevant to consider that constructing resilient structures such as the MARES partnership also facilitates addressing big questions requiring complex interdisciplinary projects which can yield key answers on resilience and therefore sustainability. Resilient partnerships addressing socio-ecological resilience pave the way to sustainability in marine science, and the use of that science to inform important decisions on food, water, and energy security around the globe. Bridging organizations such as the NOPP can add resilience to the partnering process because they provide connectivity and flexibility among the different nodes that conform to a given marine research project.

The trust needed to create these partnerships and sustain them cannot be overstated. Building an integrated science team to respond to complex inter-disciplinary solicitations is a social process built on trust and previous relationships that prioritizes common goals and that needs to set aside differences, and include everyone while creating a collaborative team spirit (e.g., Lake, 2011). This binding force, enhanced through frequent interaction and collaboration, is likely the strongest partnership bond, helps prevail in the face of programmatic, financial or scientific difficulties, and drives the participants willingness to go beyond contractual obligations.

Conclusions

Having been through a few partnering-forming processes, and being familiar with other ones, the authors believe that a large solicitation should be generally preferred to several smaller ones when inter-disciplinary marine research is to be conducted by multiple institutions over several years. Modular and integrated approaches imply more planning and coordination work for funders, but have higher chances of leading to better integration of its results, and thus to higher quality science and usability of its outcomes. The collaborative aspect of science should extend to funding agencies creating solicitations with more specific directions on science, project management, integration, and collaborations. This is an untapped intellectual resource. Funding agencies, especially those aiming at funding research that would produce societal benefits, might consider the option of using agreements and contracts as procurement vehicles to achieve those outcomes. The funder-researcher communication before and after solicitations might need to be revised and perspectives better aligned or complemented. Solicitations and request for proposals need to initiate the integration process early, when drafting those documents, and therefore avoid the unnecessary and inefficient step of conducting a-posteriori syntheses or integrations. This planning and coordination process needs to carefully select its actors to the extent possible, a task needing to balance skills, expertise, personalities and team spirit.

Early in the MARES creation process, there was initial opposition simply because the process was new and have not been used before. However, different punctual reasons and explanations over time persuaded many to move forward. It is important to keep in mind that new ideas require extra momentum/efforts and that skepticism, while necessary and important, needs to be organized to lead to constructive ideas for producing quality science.

The issue of communication, a form of connectivity enhancing resilience, was crucial in MARES, both before and after the award. Looking forward, we emphasize the importance of the whole-part-whole approach, commonly used in team sports for assembling new plays among a multitude of players. In MARES this involved communicating new approaches to the entire group, then having one-on-one discussions between the lead agency and each potential funding partner, and then integrating and communicating those individual perspectives with the group as a whole. This iterative process also enhances the resilience of the team, as those iterations facilitate adaptive assimilation and implementation of new ideas and tasks.

The role of bridging organizations such as the NOPP and others becomes fundamental in adding credibility to the partnership process as well as providing connectivity and flexibility as the neuralgic center of partnerships. Since 1997, the NOPP has launched more than 200 partnerships involving private, US Federal government, tribal, state government, academic, and international partners. Often these partnerships involve regulatory agencies who will use the outcomes of the scientific research to inform their decisions on fisheries, offshore energy, and environmental protection. In these circumstances, bridging organizations further contribute to enhancing social and/or ecological resilience as they become a key mechanism fostering adaptive governance and enabling positive interactions through enhanced connectivity that characterize efficient and effective resource management frameworks (Auad et al., 2018). Like trees, robust partnerships need roots, a trunk, branches of different sizes, and leaves to successfully host all the required social processes. This structural design, a nature-based solution, sustains the partnership while resources are accessible. Yet large partnerships, like other complex systems such as trees, can only efficiently grow to a certain limit, at which point they achieve resonable levels of sustainability and high productivity (Jørgensen et al., 2015) to flourish within limits.

Acknowledgments

We are thankful to all the people that were part of those initial conversations that inspired and supported the creation of MARES. We would like to especially thank Rodney Cluck, James J. Kendall, Beth Burkhard (Bureau of Ocean Energy Management), and Lisa Algarin (Bureau of Safety and Environmental Enforcement). We also especially thank John Farrell who, as the Executive Director of the US Arctic Research Commission brought this novel partnership approach to the attention of many. The Commission also supported international coordination and collaborations. The involvement and support of MARES by BOEM's Alaska office, Martin Jeffries, Michael Weise and Tom Drake (Office of Naval Research), by Phil McGillivary and Jonathan Berkson (US Coast Guard), as well as by Ruth Perry and Michael Macrander (Royal Dutch Shell), are greatly appreciated. Special thanks also go to the dedicated and efficient staff and leadership of the National Oceanographic Partnership Program office who provided key support and coordination among all partners. FKW is thankful for all the internal Stantec support it took to initiate and support MARES during its proposal inception and implementation, especially John Lortie, Lee Jamieson, Jeff Green, Dom Kempson, Diane Ingraham, Gerry Myers, Alexandra Eaves, Pam Neubert, Rowenna Gryba, and Cathy Finnie. GA and FKW also thank all the funding and research partners, in both the US and Canada, that made up the MARES partnership; without their knowledge, dedication, patience, and belief in our vision, none of this would have been possible. The views and opinions expressed in this chapter are those of the authors and do not necessarily reflect the official policy or position of their employers or any other agency or organization.

References

Auad, G., Blythe, J., Coffman, K., Fath, B.D., 2018. A dynamic management framework for socio-ecological system stewardship: a case study for the United States Bureau of Ocean Energy Management. J. Environ. Manag. 225, 32−45.

Jørgensen, S.E., Fath, B.D., Nielsen, S.N., Pulselli, F.M., Fiscus, D.A., Bastianoni, S., 2015. Flourishing Within Limits to Growth: Following Nature's Way. Routledge.

Lake, D.A., 2011. Why "isms" are evil: theory, epistemology, and academic sects as impediments to understanding and progress. Int. Stud. Q. 55 (2), 465−480.

Merton, R.K., 1973. The Sociology of Science: Theoretical and Empirical Investigations. University of Chicago press.

U.S. Arctic Research Commission, 2013. Report on the U.S. Arctic Research Commission Goals and Objectives for Arctic Research, for the U.S. Arctic Research Program Plan: 2013−2014, 17. http://www.arctic.gov/publications/goals/usarc_goals_2013-14.pdf.

CHAPTER 6

Partnering with the public: The Coastal Observation and Seabird Survey Team

Julia K. Parrish[a], Hillary Burgess[a], Jaqueline Lindsey[a], Lauren Divine[b], Robert Kaler[c], Scott Pearson[d] and Jane Dolliver[e]

[a]School of Aquatic and Fishery Sciences, University of Washington, Seattle, WA, United States; [b]Aleut Community of St. Paul Island Ecosystem Conservation Office, St. Paul, AK, United States; [c]Migratory Birds, Alaska Region, U.S. Fish and Wildlife Service, Anchorage, AK, United States; [d]Washington Department of Fish and Wildlife, Olympia, WA, United States; [e]College of the Environment, University of Washington, Seattle, WA, United States

Introduction

Environmental science at scale requires the collection of large data sets, encompassing the spatial and temporal extent and grain of biogeochemical and physical phenomena, as well as of physiological, behavioral, and ecological response at population and community levels (Miloslavich et al., 2018). Within marine systems, this scale ranges to a large marine ecosystem (LME) in space and to decadal oscillation over time (Auad, 2003; Blowes et al., 2019). Traditional "mainstream science" approaches to broaching these scales have involved both ship-borne science and the advent of remote sensing technologies (buoys, gliders, satellites) allowing synoptic and larger views of the world's oceans. While returning a wealth of data which can be woven into a fabric of system-level understanding (see Chapter 10), these approaches can be expensive, are not necessarily comprehensive, require highly skilled personnel to deploy and maintain, and necessitate a range of collaboration and partnership across the scientific (academic, governmental, industry, NGO) community.

Citizen science is a very different approach for overcoming the requirements of spatial and temporal data collection "at scale" wherein members of the public agree to engage with scientists in one or more facets of the scientific process, most commonly data collection (McKinley et al., 2017). Although this approach to science has been criticized as lacking in rigor, and specifically in the precision and accuracy of participants (NASEM, 2018), data-generating projects with a standardized, effort-controlled protocol; expert-led training; and pre-post training evaluation of participants, are acceptable to the scientific community (Burgess et al., 2017). This subset of citizen science projects includes both highly popular – hundreds of thousands of participants annually – and highly published – multiple peer-reviewed publications in high impact factor journals – projects, often at temporal and spatial scales measured in decades and LMEs (Theobald et al., 2015).

Marine citizen science

Rigorous, data-driven marine citizen science projects include distinct approaches to data collection (Parrish et al., 2018). *Passive participation* includes projects in which participants deploy sensors, for example on their person or their vessels. Classically, ship of opportunity programs (e.g., International Comprehensive Ocean-Atmosphere Data Set, ICOADS) are the marine exemplar, as captains have provided weather and oceanographic data along shipping routes worldwide, stretching back several centuries (Delory and Pearlman, 2018). In the modern era, volunteer-based marine sensor deployment programs include the citizens' observatory for coastal and ocean optical monitoring (Citclops), a multi-data channel sensor system to monitor water quality (Ceccaroni et al., 2020); and SmartFin, a temperature, GPS data-logger that attaches to surfboards and is deployed under the auspices of the Surfrider Foundation (www.surfrider.org/programs/smartfin).

By contrast, *active participation* programs involve people directly in the tasks (e.g., image identification, sample collection, measurement) and/or deductions (e.g., pattern recognition, species identification) of science, including *virtual* citizen science, in which the participant is interfacing entirely online, and *hands-on* citizen science, requiring the participant to be physically present in the marine environment engaged in some task.

With the advent of digital imagery, the stage was set for virtual citizen science on platforms such as the Zooniverse or Tomnod which host standardized digital image databases allowing viewers the opportunity to learn simple tasks (e.g., outline X, search for and count Y, translate Z), and perform them on as many images as they are willing to evaluate. Multiple participants can be randomly assigned to each image and a voting algorithm used to determine the likely veracity of the solution based on the number of individuals providing that solution (Fortson et al., 2012; Marshall et al., 2016). For well-subscribed platforms such as Zooniverse, tens of thousands of images can be assessed by multiple participants within days, and with no discernible differences between expert assessment and collective or "crowd" assessment (Swanson et al., 2016). Marine projects are fast becoming more prevalent in Zooniverse (14 of 83 current Zooniverse projects are marine; October 2020) and include pinniped identification from high-resolution satellite imagery (Sayer et al., 2019; LaRue et al., 2020; Wege et al., 2020), collection and processing of cetacean imagery from social media (Facebook, Twitter, Flickr; Gibson et al., 2020), and the design of sightings apps allowing uploading of location-tagged imagery for subsequent processing as in TURT (Turtles Uniting Researchers and Tourists; Baumbach et al., 2019). Multi-project platforms such as Zooniverse require a particular data structure for upload, a written protocol, a training set or tutorial for users, and some degree of feedback and near-real-time interaction with the participants to facilitate both engagement and understanding (www.zooniverse.org).

Hands-on citizen science requires that participants engage physically, often in outdoor locations. Given the realities of logistics in the marine environment, this practically means that most projects are restricted to locations where people recreate (e.g., dive sites, surf sites, public beaches) and/or live (e.g., coastal environments). Among marine projects, an exemplar is the Reef Education Environmental Foundation (REEF). Launched in 1993, REEF allows volunteer SCUBA divers and snorkelers to upload checklist data on species occurrence and abundance for marine fish populations worldwide, as well as selected invertebrate and algae species in temperate areas. Data can be categorized by participants' expertise according to species identification tests administered by REEF. Science and resource management outcomes from this project include documentation of new species (Allen et al., 2020), the incursion of invasive and other non-native species (Schofield and Akins, 2019), disease outbreaks associated with a marine heatwave (Harvell et al., 2019), the positive (Heery et al., 2018) and negative (Pate and Marshall, 2020) effects of urbanization on species abundance, and ecosystem-based modeling of habitat association (Gruss et al., 2018).

Other marine citizen science programs include both sample collection and sample-in-hand projects. Examples of sample collection include contaminant monitoring such as MusselWatch, a National Oceanic and Atmospheric Administration (NOAA) National Status and Trends water contaminant monitoring program started in 1986. MusselWatch uses contaminant loading in mussel tissue to quantify and assess spatial and temporal trends in coastal contamination, providing a baseline for anthropogenic and natural events, including chemical spills (Apeti et al., 2018). Sampling also includes marine debris programs such as the International Coastal Cleanup (ICC), a 30-year-old volunteer-based program managed by the Ocean Conservancy, involving almost a million people in over 100 countries, and generating a dataset used to assess debris loads in coastal environments (Roman et al., 2020). Finally, sample-in-hand programs are exemplified by programs monitoring biodiversity that washes up on beaches, including species from gelatinous zooplankton (GelAvista - Pires et al., 2018; Jellyfish Watch - Purcell et al., 2015) to marine birds (COASST - Parrish et al., 2017).

COASST as a case study

The Coastal Observation and Seabird Survey Team (COASST) is a 20-year-old, hands-on marine citizen science program in coastal communities ringing the northeast Pacific (Fig. 6.1). Housed at the University of Washington, COASST recruits and trains coastal residents (COASSTers) to survey a beach of their choice in a standardized, effort-controlled fashion on a monthly basis. Focal data collection projects include beachcast carcasses of marine birds and marine debris. Datasheets and verification photographs are uploaded and archived via a web-based data portal into a PostgreSQL relational database. Quality assurance/quality control (QAQC) measures include both automated completeness and within range checks, spot-checking for transcription errors, and independent verification for species identification (Parrish et al., 2018). Data are monitored in near real-time for unexpected events, which are communicated to partnering natural resource management agencies at tribal, state, and federal levels. Data are also analyzed, combined with contemporaneous atmospheric, oceanographic, and/or ecological datasets, and used to address impact questions across a spectrum of natural and anthropogenic stressors (e.g., Jones et al., 2018, 2019; Piatt et al., 2020).

Motivation to start the program

The initial motivation for COASST was the creation of a long-term, statistically documented baseline demonstrating pattern(s) inherent in beach-cast marine birds along the outer coast of Washington state, or the seasonal frequency distribution of usually observed taxa. Such a baseline allows for both subsequent investigation of the pattern itself (e.g., of the

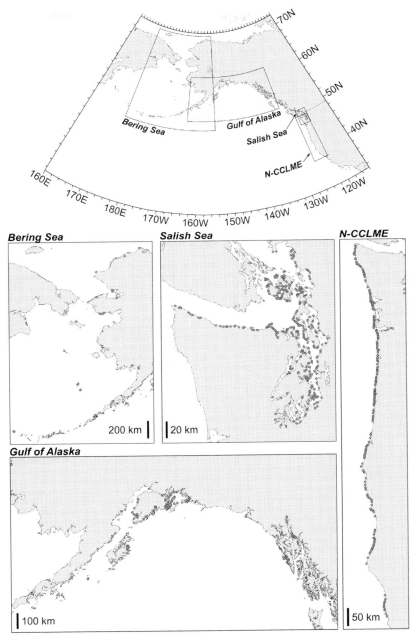

Fig. 6.1 Map of all COASST data collection sites (red dots [dark gray dots in print version]) regardless of data collection module (beached birds, marine debris), and separated into major geo-regions (*rectangles* on main map). Note the scale difference among the inset maps. Bering Sea includes sites in the Aleutian Islands, Bering Sea and Chukchi Sea. Gulf of Alaska includes sites along the Alaskan Peninsula, Gulf of Alaska and southeast Alaska. Salish Sea includes sites along the Strait of Juan de Fuca and throughout greater Puget Sound. Northern California Current large marine ecosystem (N-CCLME) includes outer coast sites in Washington, and coastal sites in Oregon and northern California.

balance between post-breeding mortality and winterkill), and documentation of departures from it or mass mortality events (Fey et al., 2015; Jones et al., 2018). Setting up COASST stemmed from one particular mortality event, the deposition of almost 4250 marine bird carcasses following the *Tenyo Maru* oil spill in July 1991, during which oil seeping from a sunken fishing vessel coated beaches from Cape Flattery at the northern tip of the coast of Washington south to Lincoln City, Oregon. At the time of this spill, there was no baseline available, which literally meant no way of assessing the impact of the oil spill on the marine bird community, and thus what amount of restoration effort would be required to rehabilitate the injured resources.

Secondary motivations quickly grew out of this core goal of a scientifically defensible baseline including attention to two feedback loops: science-education, and science-community empowerment. Increasing education outcomes at the individual level feeds back directly to the quality of the science via the rigor and longevity of the data collectors. Evaluating and documenting that participants have mastered precise data collection skills and the ability to apply accurate deductive reasoning in the species identification process, for example, is fundamental to proving the scientific worth of the ensuing information. When participants are aware of their accuracy, because of near real-time feedback from the program, they become better data collectors. When participants are also informed about the larger-scale data patterns emerging from the program as a whole, and to which they have directly contributed to, they become more engaged as well as more knowledgeable about the marine system.

Science derived from community-based participation can result in collective empowerment. In short, when science and community - defined here as a place-based, socially structured group which may also be ethnographically specific (e.g., an Indigenous community) and/or a community of practice (e.g., a fishing community) - co-develop coincident goals and pathways to reach those goals, science can directly serve the needs of the community because the community aids in and co-creates the science.

Organizational evolution

Initial years (1998–2006)

COASST was created with a grant from the David and Lucille Packard Foundation in 1998, specifically for the design, development, and pilot work to cement citizen-based data collection with an eye toward coastal marine ecology and conservation. In the development period, COASST was a small program, run by an academic Principal Investigator (a position later rebranded as Executive Director) and a postdoctoral research associate at the University of Washington, who collectively acted as jacks-of-all-trades. Once the program had secured regular data collection participation from coastal residents (2000), it expanded on campus to include an undergraduate intern program, and a Participant Coordinator (professional staff position). Beyond Packard, additional funding was secured from a range of public and private foundations beginning in 2001, mainly granted to extend and/or deepen (e.g., infill) the geographic scope of the program (Fig. 6.2).

Priorities during this initial phase included: development of a rigorous data collection protocol and QAQC measures, development of an in-person training program doable in a single day (eventually 5 h), solicitation of a stable participant corps, design of a database and linked data entry portal accessible via a website, and design of communication formats for public engagement. Almost all effort during this early period went into

Fig. 6.2 The number of participants in the COASST program. Left axis (*black line*) - cumulative over time, starting in 1999 with 12 pilot participants engaged in the development phase. Right axis (*red line [dark gray line* in print version]) - the number of participants "active" at any time, where active status is defined as having collected data within the last year. Gray overlays indicate significant impacts to participant numbers, including the geographic expansion from Washington state (first participants) into Oregon (OR), Alaska (AK) and California (CA), as well as the addition of a second data collection module focused on marine debris (MD), and the imposition of restrictions to beach access and other public health measures following the COVID-19 pandemic in the US in 2020. The horizontal axis represents time in years.

assessing and ultimately determining that the data were high quality, that data collection was sustainable within sites, and that the program could grow geographically in both grain and extent to accommodate minimum sample size and statistical power given both spatial and observer variability.

Innovations included the development of a data collection approach which disarticulated on-site carcass identification into two components: *evidence* (e.g., morphometric measurements, foot type, scaled photographs) followed by *deduction* (e.g., species identity). All data were returned to the COASST office for *independent verification* (i.e., using evidence only) of taxonomy. This approach to citizen science simultaneously facilitated the introduction of COASST data to the scientific and natural resource communities as valid, real-time, high-quality information; and the creation of participant-specific feedback when misidentifications occurred. A linchpin was the co-creation of a field key to common (60—80 species) beachcast species (Hass and Parrish, 2001) which connected the evidence collected to the deductive process of identification, and allowed for multiple pathways to determine species identification given the reality that the majority of carcasses are not intact (Bovy et al., 2016) and are thus likely to be missing characteristic body parts. COASST staff and 12 pilot participants (coastal residents of Ocean Shores, WA) engaged in a two-year adaptive iteration process to develop both the key and associated field kit (measurement tools, photographic scales, data sheets). Subsequent development of an intensive in-person training resulted in highly accurate ($\sim$70%) species identification by novice, non-experts post-training (Parrish et al., 2017).

Finally, after early attempts at recruiting and training coastal community members resulted in low (30%—50%) levels of recruitment into the long-term data collection corps, COASST developed a contractual approach, which evolved to include: a *pledge document* newly-trained individuals were asked to sign that demonstrated their commitment to survey "their beach" monthly, and a nominal deposit of ($20) for COASST kit supplies, including the field guide and printed protocol. Both of these developments had the effect of dissuading casual participation, and greatly increased the retention rate from training to first survey and beyond.

Essential to the initial years were in-kind partnerships with government agencies and community organizations that increased the number and geographic spread of data collectors to over 300 participants across four states by 2006 (Fig. 6.2). These were, in essence, outgrowths of pre-existing scientific collaborations between the COASST Executive Director and agency science and management personnel. Chief among these was the Olympic Coast National Marine Sanctuary (OCNMS) which provided a part-time position for training and participant engagement for north coast and Strait of Juan de Fuca sites in Washington; CoastWatch within the Oregon Shores Conservation Coalition which assisted in training logistics and advertising and connected COASST to a large corps of volunteers monitoring "Oregon miles" along the entire Oregon coastline; Sea Grant Marine Advisory Programs in northern California and Alaska which assisted in place-based training logistics and advertising; and the Alaska Maritime National Wildlife Refuge which incorporated COASST data collection as a base protocol for seasonal staff throughout the refuge. In all of these cases, the partnership could be described as transactional in the sense of advancing both COASST and partner goals: co-facilitating recruitment of participants or members, allowing an agency to secure additional data relevant to its management of natural resources, and/or fulfilling the mission of the agency/organization to empower local communities with respect to marine science.

Middle years (2007—2015)

Once COASST had become well established, priorities expanded from sustainable data collection to data use, and specifically to co-production of data products with tribal, state and federal partners, including peer-reviewed publications. COASST created ties to the scientific community via collaboration on analyses (e.g., Parrish et al., 2007), and effected a staffing expansion to include a Data Verifier, a Science Coordinator (both professional staff positions), and a quantitative Data Analyst (postdoctoral associate), principally funded by short-term (1—3 year), question-driven, research grants (National Science Foundation; North Pacific Research Board; Sea Grant). While the job of the Science Coordinator was to establish and maintain ties to collaborating and partnering agencies and to establish data use agreements, the focus of the data analyst was to conduct and publish analyses using COASST and partner data. Additional analysis was accomplished by graduate students incorporating COASST data into their theses or dissertations. Finally, undergraduate intern numbers grew from modest beginnings of several interns annually to 15—20 interns annually working on all facets of the program. Facilitating all of these advancements was the fact that COASST is housed at a large research university, which imparted scientific rigor to the program; provided access to students, scientific partners and grant funding opportunities; and lent credibility and stature to the program in the eyes of the public, importantly including current and future participants.

Collaboration with founding federal partner OCNMS and founding state partner Washington Department of Fish and Wildlife (WDFW) allowed COASST entry into a range of natural resource management arenas from oil spill drills to serving as the communications point-of-contact for mass mortality events. Agency partnerships also facilitated the development of indices of marine ecosystem health and other status reporting using COASST data (e.g., California Current Integrated Ecosystem Assessment, CCIEA; Harvey et al., 2019). As part of developing deeper relationships with federal resource management agencies, COASST also began working with the Alaska Fishery Observer Program and the National Seabird Program, both within NOAA, to provide mortality information on species of high fishery bycatch susceptibility. Outgrowths of this latter partnership have included the expansion of COASST field guides to include Alaskan species, and the development of a NOAA training program based on COASST field guides to increase fishery observer accuracy in monitoring marine bird bycatch.

During this phase, COASST also established loose partnerships with three other West Coast beached bird programs: BeachCOMBERS, located in central California ($34-37°N$) and operated out of the Moss Landing Marine Laboratory and Monterey Bay National Marine Sanctuary; BeachWatch, located in northern California ($37-39°N$) and operated out of the Gulf of the Farallons National Marine Sanctuary; and British Columbia Beached Bird Survey (BCBBS; $48.3-54.2°N$), operated by Bird Studies Canada, and Environment and Climate Change Canada. The four programs created interoperable datasets and began collaboration on cross-program use of data in which all organizations are explicitly acknowledged (e.g., as authors). This created a positive feedback wherein different program staff took the lead on integrated use of multi-program data to tell a science story beyond the geographic bounds of any single program. Examples include marine mammal stranding, a project spearheaded by BeachWatch scientists (Moore et al., 2009); and fishery bycatch associated beaching, a project incorporating COASST and BCBBS data (Hamel et al., 2009).

To facilitate data use, COASST established an open data policy linked to a formal data use agreement that made the conditions for use of COASST data explicit. These included: working with data users to understand their needs and to create post-processed data products where possible; and spelling out acknowledgment, attribution, ownership, and authorship requirements. The first data use agreement was signed in 2007. Since that time, COASST has fulfilled almost 250 data use requests (through Oct 2020; Fig. 6.3).

Even as COASST began to focus on data use, two secondary priorities included expansion into other data collection modules that would add value to natural resource science and management in coastal marine environments, and expansion into social science, and specifically toward understanding and researching participant motivation.

In 2013, COASST began work on a marine debris module funded by the National Science Foundation (NSF). Selection of this data stream followed from the 2011 earthquake in Tōhoku, Japan and consequent tsunami, which focused massive public attention – including from COASST participants – on beach debris. By 2016, participant data collection had started in earnest, growing quickly to 75 beaches and 150 people centered in the Salish Sea and along the coast of Oregon. Within two years of data collection, the COASST approach to debris abundance estimation and characterization had

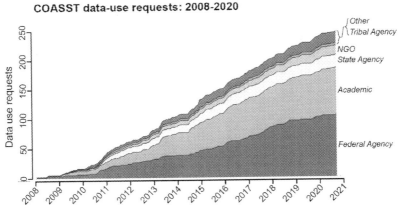

Fig. 6.3 Data use requests since January 2008 and running through October 2020, cumulative by data user category. Federal agency includes international agencies (e.g., Bird Studies Canada); state agency includes the states of California, Oregon, Washington and Alaska; NGO is non-governmental organizations; tribal agency includes sovereign nations and corporations in Washington and Alaska; other includes the news media, private individuals (e.g., book authors), and COASST participants. Data use requests include only the transfer of specific numeric data stored in the COASST database, and do not include requests for: other COASST products (e.g., field guides), services (e.g., trainings) or interviews (e.g., by the news media).

garnered support from US federal agencies (NOAA, Environmental Protection Agency), and appeared in part in national publications and protocols. This program has been highly successful, in part because of lessons learned in the design and implementation of the beached bird program. However, working within the marine debris arena presented unique challenges, including a general focus on debris clean-up versus debris science among resource management agencies, community organizations and NGOs; and a protocol that is inherently less "deduction-oriented" than the beached bird protocol (i.e., there is no "Aha!" moment on the part of data collectors as to debris identity). These issues made the development of partnerships more difficult, and the sustainability of data collectors more fragile.

Also funded by the NSF, COASST began more serious consideration of the social science of citizen science, including why and how participants learn, participant identity, and the factors leading to participant recruitment and retention. These studies revealed that the basic skills COASST participants acquire on training can be honed even further over year(s) in the program as evidenced by seasoned (one year plus) beached bird COASSTer population-level accuracy of $\sim$87% correct to species (Parrish et al., 2019). In short, longer-term participant retention improves programmatic performance. Concurrent work indicated the importance to participants of being part of a collective of people engaged in rigorous creation of the COASST dataset (Haywood et al., 2016; He et al., 2019). Additional motivations to remain active in the program included the value of the program/data to science and natural resource management (Haywood et al., 2016). These findings allowed adaptation of COASST trainings and outreach materials to forefront science and natural resource management messages (Parrish et al., 2017).

Exploring the social science of COASST necessitated the incorporation of social science professionals as program advisors and co-investigators on education research grants; thus began long partnerships with the College of Education programs at University of California Davis and Oregon State University. Extension into the realm of learning research also allowed access to additional funding resources in both public and private foundation spheres.

Both natural and social science work helped bolster the total number of active participants to over 800 by 2013 (Fig. 6.2), with concomitant expansion to the literal limits of geography: Cape Mendocino in California, where COASST abuts the CoastWatch program, and into the Chukchi Sea in Alaska (Fig. 6.1).

Finally, in 2011 COASST established an Advisory Board consisting of 20−25 volunteers including academic, agency, NGO, and community organization members; spanning natural science, social science, and education research; and including scholars, practitioners, and managers. Advisory Board functions include connecting COASST to data use, research, and funding opportunities; providing feedback on potential research directions and/or data collection expansion; and advising on long-term sustainability. Several Advisory Board members have become principal investigators, collaborators, or contractors on funded work.

Current era (2016−2020+)

In the current era, COASST programmatic priorities have focused on extending the analysis and synthesis of both natural science and social science datasets, creating community-specific programming, and creating long-term value and usefulness among partner organizations as one method of achieving sustainability.

Starting in 2015, a series of marine bird mass mortality events descended on the Northeast Pacific. Starting in the northern California Current Large Marine Ecosystem (CCLME) north throughout the Gulf of Alaska for the first two years, the phenomena extended to the Bering Sea, and then into the Chukchi and Arctic by 2016. For COASST, this period of elevated mortality allowed our emerging partnerships across the scientific community to blossom, with a series of publications melding datasets and expertise across disciplines, and university-agency boundaries (Jones et al., 2019; Piatt et al., 2020; Romano et al., 2020; Van Hemert et al., 2021). This work helped define the species and number of individuals involved, the geographic scope of the events, and the physical and ecological drivers of the mortality, increasing the long-term value of COASST and other beached bird program data in the process.

Recently, COASST has sought to create a consortium of agency-based data and service users with an interest in helping to sustain COASST data collection within their geographic and/or natural resource management purview. This approach to the long-term sustainability of citizen science data collection and curation is novel. Within COASST, this takes the forms of low-level ($10−25K per partner per annum) annual funding to:
- provide clean, post-processed data to larger multi-channel environmental databases annually (e.g., COASST-Alaska Ocean Observing System (AOOS) partnership).

- help sustain data collection sites in particular geographies, and support local data collectors (e.g., COASST-Alaska Department of Fish and Game (ADFG) partnership).
- provide a data curation and processing role in the face of continuing anomalous mortality events, including both COASST data, and additional data and information provided by other organizations, tribal governments, and individuals (e.g., COASST-U.S. Fish and Wildlife Service (USFWS) partnership).

Finally, in the last several years, our work with Indigenous communities has expanded, as COASST started to work in the Bering and Chukchi Seas and the appearance of mass mortality events throughout Alaskan waters necessitated a call to action given the rapid changes coastal communities were witnessing, including food safety and food security concerns in the face of sudden and unexplained die-offs. Here COASST also took on a secondary role as data collator, blending our regular monthly survey data with information streams including one-off surveys conducted by agency personnel through to information provided by coastal community members. Crucial to this effort was the development of modified survey techniques by the Aleut Community of St. Paul Island Ecosystem Conservation Office (ECO), which worked with COASST to co-create *Die-off Alert*, a data collection program allowing participants to collect, arrange and photograph carcasses, and text the images to COASST. Formalizing (through a Memorandum Of Understanding, MOU) our data curation role with the USFWS Alaska Regional Office also allowed COASST to maintain all beached bird data sent to the Service, and to produce overview maps of die-offs for distribution back out to communities and at other stakeholder gatherings.

Growing out of this period is a series of agreements and more formalized MOUs that define the way(s) in which COASST and the other party(ies) act and interact regarding data collection, ownership, and use. Most recently, these individual arrangements, including more longstanding ones held between COASST and tribal governments in coastal Washington, have led to developing and publishing a Code of Conduct on the COASST website (https://coasst.org/code-of-conduct/), which includes three sections: (1) within the "COASST Community" which we define as all persons participating in COASST up through collaborators and formal partners; (2) partnerships with Indigenous communities, tribal governments or tribal organizations; and (3) with individuals and organizations seeking to use COASST data.

Lessons learned
Funding
As with any long-term research project, sustaining the funding is a constant issue. The vast majority of long-lived, broad extent environmental citizen science programs are housed either within mission-focused government agencies (e.g., NOAA, NASA) or science-focused public engagement organizations with a paying membership base (e.g., California Academy of Sciences, Cornell Laboratory of Ornithology), both of which can provide continuous base operations funding.

As a university-housed program, COASST approached funding sustainability by weaving together a funding portfolio targeting: *geographic expansion* (e.g., expansion into California and Alaska) which was the dominant COASST strategy in the initial years; *programmatic expansion* (e.g., invention of additional data collection modules such as marine debris; expansion into social science research); and applying for *short-term research grants* which can rest on the existing dataset and/or program infrastructure (e.g., use of the beached bird database as a baseline to investigate taxon-specific susceptibility to oiling). The latter two approaches have become a dominant funding strategy for COASST in the present.

Creation of a consortium funding model, which intentionally taps partners interested in sustaining data collection and curation linked to open data and data uses within ecosystem health monitoring, is a third approach for citizen science, and one which may become more plausible as on-the-ground, highly rigorous, public-facing data collection programs become more numerous, and demonstrate their use and effectiveness in the public, scientific, and resource management space. A fourth funding approach links funding to provision of services and expertise which follow from core operations. In the case of COASST, this includes expert trainers (e.g., seabird identification training for fishery observers) and data analysis/visualization (e.g., near real-time mapping of mass mortality events in Alaska).

Leadership
Creating an independently funded citizen science program that is sustainable in the long-term requires a strong leader as the face-voice, champion, and street cred of the program. Someone who is accorded the stature of expert, and who can span between the worlds of mainstream science and the public in pursuit of an emergent whole. Essential to the skillset of that person are communication and networking, which result in

both expanding participation within the data collection corps, and also deepening partnerships in the relevant data user communities. Sustainability becomes, in part, an issue of whether the leader can be replaced within the larger structure of the organization housing the program. For agency, NGO and outreach organizations (e.g., museums), replacement of such an Executive Director appears more possible than within the academic model, in which early-career faculty rarely take over the research programming of senior faculty.

A secondary, and fundamental, layer of leadership comes from the COASST staff other than the Executive Director. COASST has evolved into a group of dedicated staff each responsible for one or more facets of the program: student interns, participant management, communication, science collaboration, data verification, data curation and analysis. Sustaining this approach is acknowledgment of creativity, innovation and responsibility through authorship, co-investigator status on grants and contracts, and public point of contact for program partners. However, detracting from it is the top-down nature of the academic model, which rewards faculty as the highest tier, often with sole programmatic authority.

Success and sustainability

Independent programs that cleave to a well-defined core of work while maintaining the flexibility to change with the times are best positioned for long-term success. Within COASST, this core is defined by strict adherence to the creation of a highly rigorous and relevant dataset with the spatial grain and spatio-temporal extent needed to address system-level patterns, issues, and concerns. Flexibility has been defined principally through an evolution of realizations about the needs, agency, expertise, and rights of the participant corps, and more broadly of the communities of which they are a part. Without belaboring the issue, we suggest this can be disarticulated into three overlapping points:

Communication

A system of personalized feedback, so that participants know what they are doing right, and wrong, allows individuals to hone skills and begin to teach others. Programmatic communication, from real-time alerts to research findings, allows participants to place their own hands-on experience in a larger space-time context and understand the system in ways that may not be otherwise accessible. When combined, the personalized and the programmatic can provide participants with the motivation to continue, and the agency to communicate to others including decision-makers (Haywood et al., 2016).

Listening
A good idea can come from anyone, anywhere. Although mainstream science favors a narrow definition of expert, rigorous citizen science allows participants to contribute substantively to the program in many ways. Within COASST, participants have been fundamental to training module development and testing, have contributed several toolset innovations (e.g., date-place slates now de rigueur in all photographs), and even proposed new data collection processes (e.g., *Die-off Alert*). COASSTers and community members have also sparked entire lines of scientific inquiry through effective early warning efforts (e.g., scoter die-off along the Washington coast - Jones et al., 2017; tufted puffin die-off on St. Paul, Jones et al., 2019). Moving from a one-way communication structure - science expert to public recipient - to a two-way communication feedback which recognizes place-based and community-based forms of expertise (CEOSE, 2019) is inherent in rigorous citizen science, and is gaining ground within mainstream science.

Science as a team sport
Although academia creates highly trained individuals accorded the status of "scientist" and credentialed with a Ph.D., these professionals actually work with teams of individuals - students, postdocs, technicians, and others - to affect the work of science. Assimilation of on-the-beach participants as full members of the science team, accorded with rights and responsibilities, also means that these individuals should be given full credit for their work, in addition to other forms of compensation. Citizen science is a data collection method, but it is also a social contract. In this sense, individuals are agreeing to participate in return for payment; it is simply that the currencies of the contract are not usually monetary. While academic currencies include grant leadership and authorship on peer-reviewed publications, participant currencies may also include place-based motivations (Haywood et al., 2021); and support of science, conservation, and the greater good (Haywood et al., 2016; He et al., 2019). For communities of practice where data collection speaks directly to livelihood, (e.g., fishery communities), or Indigenous communities exerting sovereign rights over data collected within their geographic

purview, and/or information culturally or historically held; participation in a citizen science program should be accorded with full data ownership and use rights which are dictated not by the program, but by a mutually crafted, formal agreement (e.g., an MOU). Increasingly, citizen science programs are adopting Codes of Conduct that specify these rights and responsibilities of all participants and programs (https://www.citizenscience.org/get-involved/working-groups/ethics-working-group/resources/) and COASST is no exception.

Conclusion - partnerships in citizen science

The COASST experience shows that the most important partnership in citizen science is between the program and the slice of public engaged to conduct the on-the-ground work. Almost as important are the partnerships allowing data flow to and use within the realms of science and resource management. Thus citizen science programs with a goal of collecting high-quality data at a scale, grain, and longevity needed to address questions and issues within environmental science can become boundary spanners (sensu Harris et al., 2021) between mainstream science and the public by partnering with both sides to simultaneously advance knowledge and science/environmental literacy. While some within the formal science and education community may view this as a reciprocal set of one-way flows - teaching/training to the public, work/data to science - a more nuanced view is of a network connecting science and the public, with multi-directional flows of information collectively furthering actionable science (e.g., Day, 2017).

Acknowledgments

During the construction of this chapter COASST staff were funded by the National Science Foundation EHR/DRL 1322820 and 2031884; Washington Sea Grant R/RCE-9; and partners within the COASST Consortium (https://coasst.org/about/our-supporters/). Parrish was additionally supported by the Wakefield family. T. Jones produced the figures. Nothing in this chapter would be possible without the tireless persistence of thousands of COASST participants who have rigorously searched beaches for birds and trash for over 20 years.

References

Allen, G.R., Erdmann, M.V., Mondong, M.U., 2020. Heteroconger guttatus, a new species of garden eel (Pisces: Congridae: Heterocongrinae) from West Papua, Indonesia. J. Ocean Sci. Found. 35, 8–17.

Apeti, D.A., Wirth, E.F., Leight, A., Mason, A.L., Pisarski, E., 2018. An Assessment of Contaminants of Emerging Concern in Chesapeake Bay, MD and Charleston Harbor, SC.
Auad, G., 2003. Interdecadal dynamics of the north pacific ocean. J. Phys. Oceanogr. 33 (12), 2483–2503.
Baumbach, D.S., Anger, E.C., Collado, N.A., Dunbar, S.G., 2019. Identifying sea turtle home ranges utilizing citizen-science data from novel web-based and smartphone GIS Applications. Chelonian Conserv. Biol. 18 (2), 133–144.
Blowes, S.A., Supp, S.R., Antão, L.H., Bates, A., Bruelheide, H., Chase, J.M., Moyes, F., Magurran, A., McGill, B., Myers-Smith, I.H., Winter, M., 2019. The geography of biodiversity change in marine and terrestrial assemblages. Science 366 (6463), 339–345.
Bovy, K.M., Watson, J.E., Dolliver, J., Parrish, J.K., 2016. Distinguishing offshore bird hunting from beach scavenging in archaeological contexts: the value of modern beach surveys. J. Archaeol. Sci. 70, 35–47.
Burgess, H.K., DeBey, L.B., Froehlich, H.E., Schmidt, N., Theobald, E.J., Ettinger, A.K., HilleRisLambers, J., Tewksbury, J., Parrish, J.K., 2017. The science of citizen science: exploring barriers to use as a primary research tool. Biol. Conserv. 208, 113–120.
Ceccaroni, L., Piera, J., Wernand, M.R., Zielinski, O., Busch, J.A., Van Der Woerd, H.J., Bardaji, R., Friedrichs, A., Novoa, S., Thijsse, P., Velickovski, F., 2020. Citclops: a next-generation sensor system for the monitoring of natural waters and a citizens' observatory for the assessment of ecosystems' status. PLoS One 15 (3), e0230084.
CEOSE, 2019. The Committee on Equal Opportunities in Science and Engineering. Biennial Report to Congress 2017-2018. National Science Foundation.
Day, J.C., 2017. Effective public participation is fundamental for marine conservation—lessons from a large-scale MPA. Coast. Manag. 45 (6), 470–486.
Delory, E., Pearlman, J. (Eds.), 2018. Challenges and Innovations in Ocean In Situ Sensors: Measuring Inner Ocean Processes and Health in the Digital Age. Elsevier.
Fey, S.B., Siepielski, A.M., Nusslé, S., Cervantes-Yoshida, K., Hwan, J.L., Huber, E.R., Fey, M.J., Catenazzi, A., Carlson, S.M., 2015. Recent shifts in the occurrence, cause, and magnitude of animal mass mortality events. Proc. Natl. Acad. Sci. U.S.A. 112 (4), 1083–1088.
Fortson, L., Masters, K., Nichol, R., Edmondson, E.M., Lintott, C., Raddick, J., Wallin, J., 2012. Galaxy Zoo. Advances in Machine Learning and Data Mining for Astronomy, pp. 213–236, 2012.
Gibson, C.E., Williams, D., Dunlop, R., Beck, S., 2020. Using social media as a cost-effective resource in the photo-identification of a coastal bottlenose dolphin community. Aquat. Conserv. Mar. Freshw. Ecosyst. 30 (8), 1702–1710.
Grüss, A., Chagaris, D.D., Babcock, E.A., Tarnecki, J.H., 2018. Assisting ecosystem-based fisheries management efforts using a comprehensive survey database, a large environmental database, and generalized additive models. Mar. Coast. Fish. 10 (1), 40–70.
Hamel, N.J., Burger, A.E., Charleton, K., Dadvison, P., Lee, S., Bertram, D.F., Parrish, J.K., 2009. Symposium Paper: bycatch and beached birds: assessing mortality impacts in coastal net fisheries using marine bird strandings. Mar. Ornithol. 37, 41–60.
Harris, L., Garza, C., Hatch, M., Parrish, J.K., Posselt, J., Alvarez, J., Davidson, E., Eckert, G., Grimes, K., Garcia, J., Haacker, R., Horner-Devine, M.C., Johnson, A., Lemus, J., Prakash, A., Thompson, L., Vitousek, P., Martin, M.P., Reyes, K., 2021. Equitable exchange: a framework for diversity and inclusion in the geosciences. AGU Advances, 2:e2020AV000359.

Harvell, C.D., Montecino-Latorre, D., Caldwell, J.M., Burt, J.M., Bosley, K., Keller, A., Heron, S.F., Salomon, A.K., Lee, L., Pontier, O., Pattengill-Semmens, C., 2019. Disease epidemic and a marine heat wave are associated with the continental-scale collapse of a pivotal predator (*Pycnopodia helianthoides*). Sci. Adv. 5 (1), eaau7042.

Harvey, C.J., Garfield, N., Williams, G.D., Tolimieri, N., Schroeder, I., Andrews, K.S., Barnas, K., Bjorkstedt, E.P., Bograd, S.J., Brodeur, R.D., Burke, B.J., 2019. Ecosystem Status Report of the california Current for 2019: A Summary of Ecosystem Indicators Compiled by the california Current Integrated Ecosystem Assessment Team (CCIEA).

Hass, T., Parrish, J.K., 2001. Beached Birds: A COASST Field Guide. Wavefall Press, Seattle, WA, p. 155.

Haywood, B.K., Parrish, J.K., Dolliver, J., 2016. Place-based and data-rich citizen science as a precursor for conservation action. Conserv. Biol. 30 (3), 476–486.

Haywood, B.K., Parrish, J.K., He, Y., 2021. Shapeshifting attachment: exploring multi-dimensional people–place bonds in place-based citizen science. People Nat. 3 (1), 51–65.

He, Y., Parrish, J.K., Rowe, S., Jones, T., 2019. Evolving interest and sense of self in an environmental citizen science program. Ecol. Soc. 24 (2).

Heery, E.C., Olsen, A.Y., Feist, B.E., Sebens, K.P., 2018. Urbanization-related distribution patterns and habitat-use by the marine mesopredator, giant Pacific octopus (*Enteroctopus dofleini*). Urban Ecosyst. 21 (4), 707–719.

Jones, T., Parrish, J.K., Punt, A.E., Trainer, V.L., Kudela, R., Lang, J., Brancato, M.S., Odell, A., Hickey, B., 2017. Mass mortality of marine birds in the Northeast Pacific caused by Akashiwo sanguinea. Mar. Ecol. Prog. Ser. 579, 111–127.

Jones, T., Divine, L.M., Renner, H., Knowles, S., Lefebvre, K.A., Burgess, H.K., Wright, C., Parrish, J.K., 2019. Unusual mortality of Tufted puffins (*Fratercula cirrhata*) in the eastern Bering Sea. PLoS One 14 (5), e0216532.

Jones, T.J., Parrish, J.K., Peterson, W.T., Bjorkstedt, E.P., Bond, N.A., Ballance, L.T., Bowes, V., Hipfner, J.M., Burgess, H.K., Dolliver, J.E., Lindquist, K., Lindsey, J., Nevins, H.M., Robertson, R.R., Roletto, J., Wilson, L., Joyce, T., Harvey, J., 2018. Massive mortality of a planktivorous seabird in response to a marine heatwave. Geophys. Res. Lett. 45 (7), 3193–3202.

LaRue, M.A., Ainley, D.G., Pennycook, J., Stamatiou, K., Salas, L., Nur, N., Stammerjohn, S., Barrington, L., 2020. Engaging 'the crowd' in remote sensing to learn about habitat affinity of the Weddell seal in Antarctica. Rem. Sens. Ecol. Conserv. 6 (1), 70–78.

Marshall, P.J., Verma, A., More, A., Davis, C.P., More, S., Kapadia, A., Parrish, M., Snyder, C., Wilcox, J., Baeten, E., Macmillan, C., 2016. SPACE WARPS–I. Crowdsourcing the discovery of gravitational lenses. Mon. Not. Roy. Astron. Soc. 455 (2), 1171–1190.

McKinley, D.C., Miller-Rushing, A.J., Ballard, H.L., Bonney, R., Brown, H., Cook-Patton, S.C., Evans, D.M., French, R.A., Parrish, J.K., Phillips, T.B., Ryan, S.F., 2017. Citizen science can improve conservation science, natural resource management, and environmental protection. Biol. Conserv. 208, 15–28.

Miloslavich, P., Bax, N.J., Simmons, S.E., Klein, E., Appeltans, W., Aburto-Oropeza, O., Garcia, M.A., Batten, S.D., Benedetti-Cecchi, L., Checkley Jr., D.M., Shiba, S., Duffy, J.E., Dunn, D.C., Fisher, A., Gunn, J., Kudela, R., Marsac, F., Muler-Karger, F.E., Obura, D., Shin, Y.-J., 2018. Essential ocean variables for global sustained observations of biodiversity and ecosystem changes. Glob. Change Biol. 24 (6), 2416–2433.

Moore, E., Lyday, S., Roletto, J., Litle, K., Parrish, J.K., Nevins, H., Harvey, J., Mortenson, J., Greig, D., Piazza, M., Hermance, A., 2009. Entanglements of marine mammals and seabirds in central California and the north-west coast of the United States 2001–2005. Mar. Pollut. Bull. 58 (7), 1045–1051.

National Academies of Sciences, Engineering, and Medicine, 2018. Learning through Citizen Science: Enhancing Opportunities by Design.

Parrish, J.K., Bond, N., Nevins, H., Mantua, N., Loeffel, R., Peterson, W.T., Harvey, J.T., 2007. Beached birds and physical forcing in the California Current System. Mar. Ecol. Prog. Ser. 352, 275–288.

Parrish, J.K., Burgess, H., Weltzin, J.F., Fortson, L., Wiggins, A., Simmons, B., 2018. Exposing the science in citizen science: fitness to purpose and intentional design. Integr. Comp. Biol. 58 (1), 150–160.

Parrish, J.K., Jones, T., Burgess, H.K., He, Y., Fortson, L., Cavalier, D., 2019. Hoping for optimality or designing for inclusion: persistence, learning, and the social network of citizen science. Proc. Natl. Acad. Sci. U.S.A. 116 (6), 1894–1901.

Parrish, J.K., Litle, K., Dolliver, J., Hass, T., Burgess, H.K., Frost, E., Wright, C.W., Jones, T., 2017. Defining the Baseline and Tracking Change in Seabird Populations: The Coastal Observation and Seabird Survey Team (COASST). Citizen Science for Coastal and Marine Conservation. Routledge, London, UK, pp. 19–38. https://doi.org/10.4324/9781315638966-2.

Pate, J.H., Marshall, A.D., 2020. Urban manta rays: potential manta ray nursery habitat along a highly developed Florida coastline. Endanger. Species Res. 43, 51–64.

Piatt, J.F., Parrish, J.K., Renner, H.M., Schoen, S.K., Jones, T.T., Arimitsu, M.L., Kuletz, K.J., Bodenstein, B., García-Reyes, M., Duerr, R.S., Corcoran, R.M., 2020. Extreme mortality and reproductive failure of common murres resulting from the northeast Pacific marine heatwave of 2014-2016. PLoS One 15 (1), e0226087.

Pires, R.F., Cordeiro, N., Dubert, J., Marraccini, A., Relvas, P., dos Santos, A., 2018. Untangling Velella velella (Cnidaria: Anthoathecatae) transport: a citizen science and oceanographic approach. Mar. Ecol. Prog. Ser. 591, 241–251.

Purcell, J.E., Milisenda, G., Rizzo, A., Carrion, S.A., Zampardi, S., Airoldi, S., Zagami, G., Guglielmo, L., Boero, F., Doyle, T.K., Piraino, S., 2015. Digestion and predation rates of zooplankton by the pleustonic hydrozoan Velella velella and widespread blooms in 2013 and 2014. J. Plankton Res. 37 (5), 1056–1067.

Roman, L., Hardesty, B.D., Leonard, G.H., Pragnell-Raasch, H., Mallos, N., Campbell, I., Wilcox, C., 2020. A global assessment of the relationship between anthropogenic debris on land and the seafloor. Environ. Pollut. 264, 114663.

Romano, M.D., Renner, H.M., Kuletz, K.J., Parrish, J.K., Jones, T., Burgess, H.K., Cushing, D.A., Causey, D., 2020. Die-offs, reproductive failure, and changing at-sea abundance of murres in the Bering and Chukchi Seas in 2018. Deep Sea Res. II 181–182, e104877.

Sayer, S., Allen, R., Hawkes, L.A., Hockley, K., Jarvis, D., Witt, M.J., 2019. Pinnipeds, people and photo identification: the implications of grey seal movements for effective management of the species. J. Mar. Biol. Assoc. U. K. 99 (5), 1221–1230.

Schofield, P.J., Akins, L., 2019. Non-native marine fishes in Florida: updated checklist, population status and early detection/rapid response. BioInvas. Record 8 (4).

Swanson, A., Kosmala, M., Lintott, C., Packer, C., 2016. A generalized approach for producing, quantifying, and validating citizen science data from wildlife images. Conserv. Biol. 30 (3), 520–531.

Theobald, E.J., Ettinger, A.K., Burgess, H.K., DeBey, L.B., Schmidt, N.R., Froehlich, H.E., Wagner, C., HilleRisLambers, J., Tewksbury, J., Harsch, M.A., Parrish, J.K., 2015. Global change and local solutions: tapping the unrealized potential of citizen science for biodiversity research. Biol. Conserv. 181, 236–244.

Van Hemert, C., Dusek, R.J., Smith, M.M., Kaler, R., Sheffield, G., Divine, L.M., Kuletz, K.J., Knowles, S., Lankton, J.S., Hardison, D.R., Litaker, R.W., 2021. Investigation of algal toxins in a multispecies seabird die-off in the bering and Chukchi Seas. J. Wildl. Dis. 57 (2), 399–407.

Wege, M., Salas, L., LaRue, M., 2020. Citizen science and habitat modelling facilitates conservation planning for crabeater seals in the Weddell Sea. Divers. Distrib. 26 (10), 1291–1304.

CHAPTER 7

Long-term sustainability of ecological monitoring: perspectives from the Multi-Agency Rocky Intertidal Network

Lisa Gilbane[a], Richard F. Ambrose[b], Jennifer L. Burnaford[c], Mary Elaine Helix[d], C. Melissa Miner[e], Steven Murray[f], Kathleen M. Sullivan[g] and Stephen G. Whitaker[h]

[a]Bureau of Ocean Energy Management, Camarillo, CA, United States; [b]University of California, Los Angeles, CA, United States; [c]California State University, Fullerton, CA, United States; [d]Bureau of Ocean Energy Management (Retired), Camarillo, CA, United States; [e]University of California, Santa Cruz, CA, United States; [f]California State University, Fullerton (Emeritus), CA, United States; [g]California State University, Los Angeles, CA, United States; [h]National Park Service, Channel Islands, Ventura, CA, United States

Introduction

The Multi-Agency Rocky Intertidal Network (MARINe) is a collaboration among academic, private, and government groups. MARINe's mission is to develop publicly accessible and scientifically robust assessments of the health of rocky intertidal habitats by collecting coordinated long-term biological survey data over a large spatial scale. The network's goal is for these data to be used to inform science-based management decisions and provide ecological context for understanding the natural and anthropogenic changes in rocky intertidal communities. MARINe's sustainability hinges on the integration of "applied" and "basic" science, flexibility to accommodate partners with different capacities and goals for data collection, and collaborative approaches to decision-making, data sharing, and funding. Thirty years after its founding, with nearly 200 study sites along the North American Pacific coast (Fig. 7.1), the network is accomplishing its mission, as its data are widely utilized in state and federal management efforts and decisions involving rocky intertidal habitats. In addition, the network continues to grow, and now includes more than 40 partner groups, making it a rarity among ecological monitoring programs.

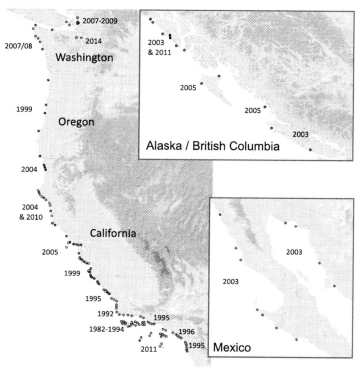

Fig. 7.1 MARINe survey sites. Site marker colors indicate different MARINe partners. Initial survey years are listed for each region.

The development and maintenance of a culture of care (e.g., Puig de la Bellacasa, 2011, 2012; Box 7.1) have been instrumental to MARINe's success and longevity. From the beginning, MARINe partners were dedicated to building a collaborative network of diverse participants to ensure the program's utility and sustainability. These early participants developed an intentional communication strategy to build and strengthen relationships among existing partners and facilitate outreach to potential new partners. United by a shared raison d'etre, scientists and stewards have worked consciously to foster deep ties among network participants and to consistently increase the relevance, functionality, and applicability of MARINe data assets. The culture of care framework grew from efforts to develop scientifically rigorous practices for data collection and data management, and is the product of brainstorming and evaluation sessions, analytical reflection, and input from the full set of participants. This culture of care operates thoughtfully and purposefully in several interdependent

> **Box 7.1 Culture of care background**
>
> In cultures of care, practices are undertaken with an awareness of and attention to the ethical implications engendered by the practices themselves, an attention that goes beyond individual intention, and instead grapples with the larger social and institutional contexts of those practices. The notion of a culture of care is increasingly applied to analyses of environmental relations, including relations between humans and the more-than-human world (Puig de la Bellacasa, 2012, 2015; Franciscus, 2015; Lien, 2015; Schrader, 2015; Bocci, 2017; Joks and Law, 2017). Puig de la Bellacasa advocates taking an approach that examines the nuances of scientific undertakings through the lens of a culture of care, arguing that care supports a strong commitment to action and obligation, even though historically, actual practices of caring —caring for others, for machines, for nonhuman species— have largely been deemed socially less valuable than many other kinds of social practices (2011:92—94; also Martin et al., 2015:628). Feminists have challenged the notion that knowledge production is separate from the cultural practices, including care, that form the context for knowledge production (Haraway, 1988; Puig de la Bellacasa, 2011, 2012; Code, 2015; Martin et al., 2015). General operational definitions for a culture of care exist (for example see Klein and Bayne, 2007; Mensik et al., 2019). However, generalization is fraught because all cultures are context-dependent, historically specific, adaptable, and very often, socially hierarchized. Cultures of care thus vary greatly, evincing different norms, different purposes, and different outcomes (Levine, 2003; Friese, 2013; Atkinson-Graham et al., 2015; Martin et al., 2015; Murphy, 2015; Bocci, 2017; English-Lueck and Avery, 2017; Zee, 2017). This chapter examines the sustainable culture of care that MARINe has developed over the last several decades.

registers: (1) rigorous practices of taxonomic identification and data collection; (2) data preservation and accessibility [to ensure continuity over time and among sampling groups]; (3) fair and inclusive governance and management, with an emphasis on consensus that respects the diverse needs of the partners; and (4) science-based decision-making. This has resulted in a program with a positive feedback loop of success; the more frequently MARINe data are used to address important questions, the more motivated new groups are to join, enhance, and expand MARINe's efforts. This chapter examines the key elements of MARINe's sustainability, with a focus on the specific practices and lessons learned that have led to the long-term and large-scale success of this network.

Foundations of MARINe
The setting for MARINe

Ecological monitoring is a critical component of ecosystem management because it increases scientific understanding of natural systems, which informs managers when creating and evaluating policies. In the 1970s, a small group of university researchers and government agency scientists located in California recognized the value of scientific monitoring for capturing long-term ecological trends in rocky intertidal biological communities. This recognition, coupled with an increased awareness of the value of long-term ecological datasets, set the stage for the formation of the monitoring program now known as MARINe.

In 1975, prior to an offshore oil and gas lease sale off southern California, the Bureau of Land Management (BLM) conducted three years of intensive, interdisciplinary baseline studies within key marine habitats. Many of the scientists who later became involved in MARINe were connected to these studies, either as principal investigators, independent researchers, or students. Although subsequent monitoring was an initial goal of this BLM program, funding ended because of a general shift within the scientific community to prioritize experimental studies. However, the value of long-term monitoring was reaffirmed in 1989 after the catastrophic *Exxon Valdez* oil spill in Alaska (Box 7.2), where a lack of baseline data severely limited the ability to determine the extent of ecological damage. The *Exxon Valdez* event spurred the Minerals Management Service (MMS) to inventory the shoreline habitats in Santa Barbara County, an area vulnerable to impact from oil and gas extraction activities. This Shoreline Inventory Project was advised by a Scientific Advisory Panel which included rocky intertidal principal investigators from the earlier BLM study and a leading scientist with the National Park Service (NPS). By that point, the NPS had monitored rocky shore (intertidal) biota on the nearby Channel Islands for nearly a decade using protocols that became the basis for MARINe's standardized protocols.

As a result of the Shoreline Inventory Project, an informal partnership was formed between MMS, the University of California, Santa Barbara (UCSB), the County of Santa Barbara, and the NPS. This partnership later evolved into MARINe. Each partner brought strengths and unique goals to the program. Investment by MMS was based on its need to collect data that could be used to address questions related to oil and gas operations in US Federal waters. Participation by UCSB and other academic researchers

Box 7.2 Timeline of events key to the formation and expansion of MARINe. For a more detailed timeline see pacificrockyintertidal.org (https://marine.ucsc.edu/ovemew/history/index.html)

1975	BLM conducts three years of intensive, interdisciplinary baseline studies within key marine habitats. Many of the scientists later were involved in MARINe.
1982	The National Park Service begins monitoring (Channel Islands, California).
1989	*Exxon Valdez* tanker oil spill (Alaska).
1990	Four partners begin monitoring in Santa Barbara County (California) under the project name: Shoreline Inventory Project.
1997	**Formal start of MARINe at the "Interagency Rocky Intertidal Monitoring Workshop" with 12 funding partners (Santa Barbara, California).**
1999	MARINe develops a shared database.
	Partnership for Interdisciplinary Studies of Coastal Oceans (PISCO) adds biodiversity sampling protocol and expands MARINe's spatial surveys into Canada and Mexico.
2000- present	Examples of MARINe data uses:
	• California Marine Protected Area design and assessment;
	• Rapid response and injury assessment for four coastal oil spills;
	• Data and expertise support for the US Federal listing of black abalone as an endangered species; and
	• Sea star wasting disease detection and tracking along US west coast.

advanced their interest in rocky intertidal communities and ensured that protocols were scientifically rigorous and statistically sound. These scientists also provided a cadre of students and volunteers with skills needed to perform the fieldwork and data-related tasks. The County of Santa Barbara was interested in tracking ecological trends and habitat changes for use in planning and permitting. The NPS brought a long view to the program, consistent with its mission, and a perspective on resource protection that valued partnerships and supported a culture of care.

Convinced that other agencies (especially the California Oil Spill Prevention and Response [OSPR], newly formed under the Oil Pollution Act of 1990) would benefit from shoreline monitoring data, MMS and

OSPR held a special session at the 1995 Southern California Academy of Sciences Meeting to discuss the need for long-term monitoring. By then, cultural shifts in government agencies and among ecological scientists were conducive to the creation of a sustainable long-term monitoring program. At this meeting, there was a groundswell of support for a long-term monitoring project with the recognition that sustained data collection could best be accomplished through collaboration.

MARINe's beginning and structure

The early efforts described above laid the groundwork for the "official" formation of MARINe in 1997, at a workshop where representatives from 12 local, state, and federal agencies, private organizations, and universities agreed to provide financial support for the program. At this workshop four basic requirements for membership were identified: (1) engagement in discussions on program improvement, (2) contribution of in-kind or monetary support to the program, (3) submission of collected data to a common database, and (4) joint publication of findings. Agreeing that "emphasis on common rather than individual goals" is beneficial to science (Fang and Casadevall, 2015), the group made the decision to monitor 55 rocky intertidal sites along the California coast using collaboratively developed protocols and a shared database.

A MARINe Steering Committee was formed after the 1997 workshop to compose a mission statement and identify primary goals, a process that helped set the focus and collaborative tone for the partnership. MARINe's first mission statement was: "To determine the health of the rocky intertidal habitat and make this information available to the public." While the concept of "health" and how to quantify it was controversial then and remains challenging for academics and decision-makers (Murray et al., 2016), the term captured the essence of key questions faced by funding agencies. Agreement on an overarching mission and early emphasis on specific goals helped to focus the growth of the partnership and to ensure that funding was concentrated on the most important tasks. A Science Panel and Database Panel were formed to work with monitoring teams in advancing shared goals (Fig. 7.2). The Database Panel's first responsibility was to propose database design options to the Steering Committee. A robust yet simple database design was supported by the Steering Committee over a more complex system, expediting the significant benefit of having a data management system in place. This initial database

Long-term sustainability of ecological monitoring   115

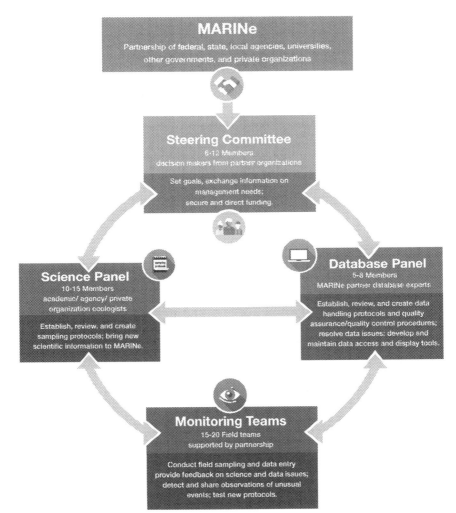

Fig. 7.2 MARINe communication structure. MARINe's goals are advanced through oversight from the Steering Committee and informed by the Science Panel, Database Panel, and Monitoring Teams.

development, including the incorporation of historic data, required five years of substantial effort but was a critical piece of the partnership ensuring data compatibility and accessibility.

The Science Panel, drawn from a diverse set of MARINe members, initially focused on the goal of developing protocols for monitoring rocky intertidal habitats that would produce data suitable for answering a range of management questions. Using knowledge gained in the aftermath of the *Exxon Valdez* oil spill, they focused on designing a scientifically rigorous but practical program. Recognizing the value of the historic NPS data, a key component was ensuring that protocols used by all groups were standardized and "backward-compatible" with protocols used by the NPS since 1982 and by the Shoreline Inventory Project since 1992. The Science Panel recommended methods that incorporated repeated sampling at rocky intertidal sites distributed over a broad geographic area to create the capacity to detect effects of unforeseen major disturbances using a network of impacted and un-impacted reference sites. The Science Panel continues to test and revise established protocols, vet new methods and approaches, troubleshoot taxonomic issues, and conduct periodic program reviews.

A simple and inclusive approach to partnership

MARINe's simple membership requirements have resulted in sustainable and long-term commitments from many different types of organizations (public and private), North American Nations, citizen science groups, multiple universities, and local, state, and federal government agencies. When MARINe was established, large-scale collaborations among academic and government scientists were rare. The diversity of MARINe participants ensured that collected data were scientifically rigorous, as well as relevant and applicable for government (both federal and state-level) decision-makers. Academic scientists recognized the importance of providing data that could be directly integrated into management decisions. Government scientists understood the pressures on academic scientists to engage in productive research projects and supported scientific publications. The requirement to contribute monetarily or with in-kind resources ensures that each partner organization is invested in MARINe and contributes to the broad funding base essential for sustaining the program. Membership flexibility has also facilitated the continuation of MARINe over the years. There are no mandates or binding agreements; partners can be active during periods when they have funding or specific needs and can scale back their activities, if necessary, during funding gaps.

Structure is necessary in any large network. MARINe uses a consensus-based governance process where input from partners is equally weighted. Governance by consensus means that decisions take longer than with a top-down model of decision-making; changes to data collection protocols, for example, can involve years of debate before integration. However, this slow process is beneficial because MARINe's consensus-based governance results in mutual understanding of the needs of each partner, allows for consideration of the scientific implications, and enhances "buy in" from all members. The MARINe mission requires multiple groups of people to collect data in the same way over many years, which is surprisingly difficult because individuals naturally make small changes to repeated processes, sometimes called "protocol drift." The MARINe governance process reduces the risk of "protocol drift" among groups because all aspects of data collection, including protocol formation, assessment, and refinement are shared through collaborative practices. This integrative approach, in which protocols become social practices with their norms and norming processes, forms part of the MARINe culture of care.

Sustainable practices

Protocol development

The vision for MARINe at the outset was to design a program that could be implemented and sustained over many years. Both Department of Interior federal agencies supporting MARINe—the Bureau of Ocean Energy Management (BOEM, formerly MMS) and the National Park Service (NPS)—share a long-term view as stewards of the natural environment. The NPS preserves unimpaired natural spaces for present and future generations, and BOEM monitors the environment adjacent to offshore energy operations. Thus, the risk of an oil spill was a primary concern for both agencies. As such, the need for long term, standardized, sustainable protocols dominated early MARINe discussions. MARINe protocols needed to be statistically robust to assess impacts from oil spills, yet simple enough that volunteers could be trained to carry them out. There was also a desire to ensure compatibility with the NPS Channel Islands rocky intertidal monitoring program (Davis 1993, 2005). A comprehensive study design was developed to adequately address the spatial (within- and among-site) and temporal replication that is important in the highly variable rocky intertidal habitat (Kingsford and Battershill, 1998; Murray et al., 2006). Given the differences among partners, e.g., universities with large numbers of trained scientists and students, as well as agencies with limited trained staff and resources, the development of a "core" set of protocols for use at every site became the focus.

Drawing on the work of the NPS, MARINe's core protocols were designed to sample a site in a single low tide exposure period. Utilizing fixed plots that targeted one of several "focal" intertidal species (pacificrockyintertidal.org), the core protocols are easy to implement and build explanatory power through sample replication over time (Raimondi et al., 2018). Core protocol data have been successfully used for their original purpose in the assessment of injuries resulting from four separate oil spill events (Raimondi et al., 2009, 2011). These protocols have also proved effective for unforeseen applications including MARINe's early detection of the decline of the black abalone (Richards and Davis, 1993; Altstatt et al., 1996), which led to its inclusion on the US Federal endangered species list. Annual sampling using core, focal species assemblage protocols is complemented by another set of protocols designed to more thoroughly capture the biodiversity, abundance, and distribution of species throughout a site. Because of the exceptional taxonomic knowledge required for these surveys, they are repeated on a less frequent basis (typically every 5—7 years) than core protocols and depend on a specialized team of researchers with specifically secured funding. This mixed design, combining frequent, less complex, fixed-plot surveys with less frequent but more comprehensive surveys enabled MARINe to overcome the challenges encountered by most field research programs in maintaining spatially and temporally expansive, yet taxonomically comprehensive surveys (Estes et al., 2018).

An unusually high degree of data consistency for such a large, multi-decadal network was achieved through the culture of care in MARINe, which is centered on a commitment to the long-term, and serves as an important asset for documenting change within rocky intertidal communities. MARINe's culture of care surrounding protocols extends to robust evaluation and adaptive management processes, again based on consensus governance. Protocols are rigorously evaluated to enhance the value of monitoring efforts and ensure that resulting data will be useful for making comparisons over broad temporal and geographic scales. Any partner group that identifies a specific regional need or a network-wide data gap is encouraged to develop supplemental protocols. For example, intertidal water temperature data recorders are deployed using one protocol by approximately only half of MARINe groups, where funding levels can support this additional effort. During development, supplemental protocols are tested by a group for at least two years. All partners in MARINe then review the results to determine if the additional protocol or new technology has merit. Following additional refinement and peer-review, the protocol may be accepted as a supplement to core protocols on a network-wide basis. Likewise, existing practices are also regularly reviewed and discussed by the membership to ensure continuity in addressing core goals and to facilitate the adoption of new technology. This collaboration enhances the quality of the scientific effort, as well as the sustainability of the organization by socializing all participants in MARINe practices.

Shared database

The early adoption of a standard network-wide database was crucial to the long-term utility of MARINe data and promoted cohesiveness among groups. MARINe's first database was relatively simple in design but sacrificing some complexity early on allowed for more rapid implementation, which was key to achieving the goals of a network-wide data management system: consistency, compatibility and accessibility. After nearly 15 years, MARINe data were migrated to a more complex database that was enhanced by ideas compiled over many years for incorporating better error prevention and data quality assurance features and increasing flexibility for incorporating individual groups' needs.

Investment in MARINe's database has fluctuated over the last 20 years and remains a challenge. While a database is most expensive during the "building" period (i.e., developing the database structure, data entry forms, error checking process), substantial and sustained support is necessary to keep the system updated and to incorporate new data. Recognizing this ongoing expense, as well as the tremendous value a functioning database brings to a program, allows group members to factor this key component of long-term monitoring into new funding opportunities. One major benefit of a partnership approach is that members can share this cost, and thus, long-term sustainability is not threatened if the contribution from any one group fluctuates over time.

A standardized, shared database is an essential component for ensuring continuity and resilience within an organization because partners may come and go, but data are retained, and analyses can continue. In addition, a well-functioning database increases the potential to attract new members for two key reasons: (1) new groups do not have to develop a data management system but can instead plug into one that is already functioning, and (2) data from a single site or handful of sites can be examined within a much broader geographical context (e.g., asking if trends observed at a local site are happening at a regional scale). Also, many funding agencies recognize the importance of data availability and have increasingly stringent requirements for how data are stored and made available. For all the reasons outlined above, the resilience of the MARINe partnership has been enhanced by its thoughtful approach to data management.

Communication

Communication has been key to the longevity of MARINe. The organization's communication strategy prioritizes a core set of critical components to enable groups to participate over decades, even when funding and staff are minimal. In this section, we focus on two primary components: an annual meeting and our website. Both have been integral to the evolution and persistence of MARINe and provide important avenues for exchange of ideas within the network and with the public and other interested groups.

In 1997, 50 people gathered at the "Interagency Rocky Intertidal Monitoring Workshop", to ensure methodical consistency across the MARINe network. Since this inaugural event, the annual in-person MARINe meeting has been essential for the maintenance of a productive and cohesive network. The meeting is designed to encourage partners to share experiences and discuss needs, provide a platform to identify next steps, train partner teams, and standardize protocols and species identifications. All members of each partner group (e.g., students, field teams, team leaders, principal investigators, funders) are encouraged to attend and actively engage in the meeting. Moving the location facilitates engagement, as groups with limited travel funds can send more representatives to "local" meetings than to "distant" meetings. This emphasis on inclusion deepens the investment of partner groups by creating a sense of belonging. The development of trust and respect through in-person interactions allows successful engagement and resolution of tough issues throughout the year. Robust moderated discussions allow attendees to make decisions and act with deep and broad support from partners. The meeting is critical to MARINe's sustainability because clear communication among groups limits misunderstandings and misinterpretations, which is important given the very different concerns, pressures, and constraints for federal, state, non-profit, and academic partners.

MARINe's commitment to a cost-effective, inclusive meeting with rotating geographic location also facilitates outreach. Including a diverse set of participants in every meeting ensures that the network remains visible and relevant and helps the network to grow. Attendance by network partner personnel beyond each group's core field or technology team increases buy-in. For example, government agency supervisors who attend a local meeting gain awareness of the benefits of continued investment in the network and in the value of long-term monitoring programs. Stakeholder groups invited to attend local meetings learn about the network and often become partners. Through careful planning and commitment, the MARINe annual meeting brings together field teams, funding representatives, supervisors, and newly interested parties from government agencies, non-profits, and academic institutions along the entire coast, and sets the table for on-going, growing, and productive collaboration.

MARINe's website is another core component of the communication strategy, as an open, flexible, and accessible platform for dissemination of information about methods, data, and summaries of our findings. One of the network's founding principles, which set the group apart from similar efforts in the 1990s, was that data would be made publicly available. The MARINe website is essential for achieving this goal because it is: (1) regularly updated to include current protocols, up-to-date data summaries, and findings within the network; (2) easily viewed by everyone, including government partners who are often restricted by the types of platforms they can access; and (3) includes geospatial tools for data exploration and display. MARINe's open data sharing policy has expanded data use in ways that were initially unanticipated, and as a result, the overall scientific value of the program has been strengthened. One example of this expansion is the use of MARINe data by high school and college instructors who have used original data for distance learning activities to enhance or replace field studies.

Continuity

Like any other multidecadal monitoring program, MARINe needs to ensure continuity in people, protocols, and funding to be sustainable. The institutional memory and professional experience of the many key personnel who have been part of MARINe since its inception are critical for maintaining MARINe's culture of care. This high degree of personnel retention impacts all aspects of the partnership. MARINe's database managers have years of experience in both fieldwork and data organization, which gives them exceptional insight into the links between data collection, entry, quality control, and application. Members of most field sampling teams have decades of experience with MARINe, and the network's emphasis on taxonomic training has led to their emergence as experts in the identification of US West Coast rocky intertidal species. MARINe invests in personnel and their development, which has resulted in more efficient, accurate, and consistent sampling. MARINe's experienced field samplers have identified potential procedural and/or ecological issues (e.g., detecting sea star wasting disease) and adjusted protocols as necessary. Long-term participants bring experience and perspective that helps resolve the inevitable problems or inconsistencies that arise in a long-term study with a complex data set.

MARINe values long-term continuity while also recognizing the importance of continually evaluating current protocols and objectives and incorporating new technology to improve efficiency and accuracy. For example, improved camera technology has been continuously integrated into MARINe core protocols to ensure maximal photo quality and resolution. Data collection originally used Palm Pilots equipped with barcode scanners but now uses a custom-designed app. MARINe continually assesses protocol utility through regular programmatic reviews and statistical analyses (Engle et al., 1997; Richards and Davis, 1993; Lafferty, 2001; Minchinton and Raimondi, 2001) to ensure data acquisition effectively addresses goals and objectives. For example, when funding for sampling was reduced, a statistical power analysis informed the decision of whether to reduce the number of sites or retain all sites but reduce sampling effort to once per year (Raimondi et al., 2018).

Another important component of continuity is that new data remain "backward compatible" with older data, particularly regarding species taxonomy. To ensure compatibility of taxonomic information over time, a three-pronged approach was implemented that included: (1) repeated training sessions in which members of all research teams learn or refresh their understanding of species identifications, (2) repeated collections of voucher specimens to provide long-term records of species identifications by sampling groups, and (3) establishment of long-term relationships with algal and invertebrate taxonomic experts to ensure that training sessions, and voucher specimen identification and archiving, incorporate current knowledge and up-to-date standards.

MARINe's success in retaining participants is due to both the nature of intertidal fieldwork and the culture of care fostered by the network. A passion for understanding and protecting the marine environment goes beyond a job description and contributes to members' initial involvement and career satisfaction. Equally important, though, is the intentionally inclusive community developed by MARINe, in which members feel that their contributions are key to network success. Many MARINe participants started as students and were immersed in the MARINe culture before moving onto jobs with other MARINe teams in academia or government. Because of MARINe's longevity, many of the founding members have retired; yet the leadership transitions have happened smoothly, with minimal loss of long-term program knowledge, largely because of the culture of care which supports shared knowledge and encourages participation. This intense personal commitment strongly benefits MARINe, as members are invested in finding solutions to coastal management issues with the knowledge that their findings will translate directly into more effective state and federal agency management.

Funding

A final key component of MARINe's continuity and long-term sustainability has been a strategy for keeping annual operational costs as low as possible while maintaining rigorous high-quality data collection from a diverse group of partners. The costs involved with establishing and maintaining a long-term ecological monitoring program include budgetary costs for program development in addition to the operational costs involved with scientific oversight, personnel training, data collection and management, analysis, reporting, and administration (Caughlan and Oakley, 2001). For MARINe, initial funds were focused on creating sustainable protocols and a shared database. Once operational, the costs and time involved with managing data through established data life cycles must be allocated and require 25%—30% of a program's annual budget (Caughlan and Oakley, 2001, and references therein). Meeting all these requirements is challenging and costly in the face of limited, fluctuating, or uncertain monitoring program budgets. When MARINe faced cuts, they were made according to established priorities and were made to publication efforts rather than people or database maintenance. Retaining people and the database are MARINe's ultimate assets and the foundation for a return on all invested funds.

MARINe's sustainable funding model is dependent upon a diversity of partners from different types of institutions. Government agencies, university researchers, and private organizations are the three primary

MARINe member types. Each type has different budget priorities and funding cycles, which over time, can complement one another and provide a diverse and resilient portfolio of funding options. Government funding can be challenging because the priorities and management goals of each agency can shift over time. MARINe has refocused its application of data several times to be relevant to the current needs of an agency's management and mission priorities. Government contracts and agreements are mandated to be re-advertised and some of MARINe's funding comes in one-year increments, which creates challenges for long-term planning and staff retention. The key to government funding that can weather typical proposal cycles is having knowledgeable staff within each agency articulating the benefits of funding MARINe within their organization's unique procurement culture and processes. University researchers receive funding from government agencies committed to MARINe, which can in turn, be leveraged to gain additional support from private organizations and grant-based funding sources. This approach to diversifying funding sources has enhanced and sustained MARINe's efforts and data utility. For instance, when the black abalone was added to the US Federal endangered species list, MARINe university and NPS researchers, who were already monitoring these populations, leveraged existing data to successfully apply for contracts and grants to further study black abalone. In turn, other groups funding and monitoring black abalone eventually joined MARINe as partners. Reliance on one institution or funding organization is rarely sustainable over the long-term. A diverse funding model is needed to respond to changing funding priorities and to establish a feedback loop of success that allows partnership efforts to grow.

MARINe's collaborative philosophy supports financial connections among its diverse partners that are built around a shared concern for rocky intertidal habitats. For example, during the 1997 *Torch* spill in Santa Barbara County, MARINe assisted by providing scientists from several state, federal, and local agencies. In return, the county provided monies that allowed MARINe to develop their database. During the 2007 *Cosco Busan* oil spill, MARINe supported the NPS sites affected by the spill by providing field teams, which led to increased monetary support for rocky intertidal monitoring at other NPS sites along the Pacific coast. In a culture of care, funding is more than just money. It is a vehicle for supporting collaborative monitoring and data collection efforts, the core of MARINe's mission.

Strengths of the partnership

MARINe values long-term ecological monitoring because it provides critical scientific information that natural resource managers need to make informed decisions. Resilience in long-term ecological monitoring programs comes from balancing the multiple goals of remaining as cost-effective as possible and addressing core questions, while remaining adaptable and relevant as new questions arise (Iles, 1994; Overton and Stehman, 1996). It can be difficult to maintain support for the cost involved with long-term monitoring because benefits are often not immediately perceivable and may not be expressed in monetary terms (Caughlan and Oakley, 2001). For instance, MARINe's first successful joint publication was produced five years after the partnership was formally established. However, once a partnership is well-developed, that initial investment is returned. For example, new groups joining MARINe do not have to develop their data management system but can instead plug into the MARINe database and existing data output scripts and visualizations. Significantly, a new group's initial data collection efforts at even one site have immediate, broader context within the larger network of sites, and therefore enhanced value. Thus, a relatively small effort from a small organization can contribute to spatially and temporally large questions, which is extremely cost-effective.

The benefits of a sustainable partnership have surpassed the expectations of the MARINe founding members in both the breadth (geographic scale and number of partners) and utility of the program. The value of information obtained from MARINe data has proven to be well worth the cost of monitoring due to the program's ability to accurately and consistently detect and measure changes in rocky intertidal communities. This includes the detection of impacts from "ecological surprises" that protocols were not originally designed to address, such as novel marine diseases, impacts from protection policies, and climate anomalies. MARINe data have provided resource managers and researchers with information to respond effectively to both known and emerging management issues (Richards and Davis, 1993; Altstatt et al., 1996; Raimondi et al., 2002; Miner et al., 2018).

Perhaps most importantly, MARINe as a partnership has built awareness and engagement among policymakers, managers, scientists, and the public regarding long-term changes in the composition and resilience of rocky

> **Box 7.3 MARINe's lessons learned**
>
> - Foster high retention of knowledgeable, passionate personnel and recruitment of new members through a culture of care that supports equality in member contribution;
> - Seek out partners with diverse data application goals and funding mechanisms;
> - Build data storage, maintenance, and sharing mechanisms into the program budget, including a dedicated data manager; and
> - Keep early annual operational costs as low as possible with protocols and database structure that are simple in design.

intertidal communities. MARINe's commitment to developing and maintaining a long-term, widely shared monitoring effort rests on much discussed and mutually agreed upon core values, including a delineated scientific mission, methodology, and a commitment to making scientific research applicable to immediate and long-term management efforts. MARINe has created a sustained culture of care: care for the rocky intertidal habitat vital to the health of the Pacific coastal zone, care for scientific rigor, care for collaboration, care for maintaining continuity, and care for the future. The MARINe culture of care has nurtured the perseverance necessary for sustaining a long-term ecological research network and has provided lessons (Box 7.3) and a context in which the many carefully developed practices and tools (standardized protocols, communication tools, training, and database) have become more than a mere sum of parts.

Acknowledgments

We acknowledge the hundreds of people that have contributed to the establishment and maintenance of the Multi-Agency Rocky Intertidal Network. If there is a low tide on the US west coast, there is someone using a MARINe protocol. Pete Raimondi was the keystone for growing the partnership, increasing the number and areal extent of the sites across the west coast, and enabling our data to inform Natural Resource Damage Assessments. Jack Engle was instrumental in coordinating contributions by MARINe partners and meticulously tracking changes to the MARINe protocol. We thank Steve Weisberg, Dan Richards, Mike Anderson, and John Steinbeck for their long-standing contributions and support. We thank the reviewers of this chapter, who also contribute significantly to

MARINe: Gary Davis, Pete Raimondi, and Jayson Smith. This study utilized data collected by the Multi-Agency Rocky Intertidal Network (MARINe): a long-term ecological consortium funded and supported by many groups. Please visit pacificrockyintertidal.org for a complete list of the MARINe partners responsible for monitoring and funding these data. Data management has been primarily supported by BOEM (Bureau of Ocean Energy Management), NPS (National Park Service), The David & Lucile Packard Foundation, and United States Navy. The views and opinions expressed in this chapter are those of the authors and do not necessarily reflect the official policy or position of their employers or any other agency or organization.

References

Altstatt, J.M., Ambrose, R.F., Engle, J.M., Haaker, P.L., Lafferty, K.D., Raimondi, P.T., 1996. Recent declines of black abalone Haliotis cracherodii on the mainland coast of central California. Mar. Ecol. Prog. Ser. 142, 185–192. https://doi.org/10.3354/meps142185.

Atkinson-Graham, M., Kenney, M., Ladd, K., Murray, C.M., Simmonds, E.A.J., 2015. Care in context: becoming an STS researcher. Soc. Stud. Sci. 45 (5), 738–748. https://doi.org/10.1177/0306312715600277.

Bocci, P., 2017. Tangles of care: killing goats to save tortoises on the Galápagos Islands. Cult. Anthropol. 32 (3), 424–449. https://doi.org/10.14506/ca32.3.08.

Caughlan, L., Oakley, K.L., 2001. Cost considerations for long-term ecological monitoring. Ecol. Indicat. 1 (2), 123–134. https://doi.org/10.1016/S1470-160X(01)00015-2.

Code, L., 2015. Care, concern, and advocacy: is there a place for epistemic responsibility? Feminist Phil. Q. 1 (1). https://doi.org/10.5206/fpq/2015.1.1.

Davis, G.E., 1993. Design elements of monitoring programs: the necessary ingredients for success. Environ. Monit. Assess. 26 (2), 99–105. https://doi.org/10.1007/BF00547489.

Davis, G.E., 2005. National Park stewardship and 'vital signs' monitoring: a case study from Channel Islands National Park, California. Aquat. Conserv. Mar. Freshw. Ecosyst. 15 (1), 71–89. https://doi.org/10.1002/aqc.643.

English-Lueck, J.A., Avery, M.L., 2017. Intensifying work and chasing innovation: incorporating care in Silicon Valley. Anthropol. Work. Rev. 38 (1), 40–49. https://doi.org/10.1111/awr.12111.

Engle, J.M., Ambrose, R.F., Raimondi, P.T., 1997. Synopsis of the Interagency Rocky Intertidal Monitoring Network Workshop. OCS, MMS Study 1997-0012. Coastal Research Center, Marine Science Institute, University of California, Santa Barbara, California. MMS Cooperative Agreement Number 14-35-0001-30761. https://www.boem.gov/ESPIS/3/3502.pdf. Accessed November 20, 2020.

Estes, L., Elsen, P.R., Treuer, T., Ahmed, L., Caylor, K., Chang, J., Choi, J.J., Ellis, E.C., 2018. The spatial and temporal domains of modern ecology. Nat. Ecol. Evol. 2 (5), 819–826. https://doi.org/10.1038/s41559-018-0524-4.

Fang, F.C., Casadevall, A., 2015. Competitive science: is competition ruining science? Infect. Immun. 83, 1229–1233. https://doi.org/10.1128/IAI.02939-14.

Franciscus, P., 2015. Laudato si: On care for our common home. Vatican Press, Vatican City, p. w2.

Friese, C., 2013. Realizing potential in translational medicine: The uncanny emergence of care as science. Curr. Anthropol. 54 (7), S129–S138. https://doi.org/10.1086/670805.

Haraway, D., 1988. Situated knowledges: the science question in feminism and the privilege of partial perspective. Fem. Stud. 14 (3), 575–599. https://doi.org/10.2307/3178066.

Iles, K., 1994. Directions in forest inventory. J. For. 92 (12), 12−15. https://doi.org/10.1093/jof/92.12.12.

Joks, S., Law, J., 2017. Sámi salmon, state salmon: TEK, technoscience and care. Socio. Rev. 65 (2_Suppl. l), 150−171. https://doi.org/10.1177/0081176917710428.

Kingsford, M.J., Battershill, C.N., 1998. Subtidal Habitats and Benthic Organisms of Rocky Reefs. Studying Temperate Marine Environments. A Handbook for Ecologists. A, CRC Press, Boca Raton, FL, USA, p. 84.

Klein, H.J., Bayne, K.A., 2007. Establishing a culture of care, conscience, and responsibility: addressing the improvement of scientific discovery and animal welfare through science-based performance standards. ILAR J. 48 (1), 3−11. https://doi.org/10.1093/ilar.48.1.3.

Lafferty, K.D., 2001. Channel Islands National Park Intertidal Monitoring Program Review, January 17-19, 2001. Channel Islands National Park, Unpublished Report, Ventura, California.

Levine, P., 2003. Prostitution, Race, and Politics: Policing Venereal Disease in the British Empire. Psychology Press, New York, USA.

Lien, M.E., 2015. Becoming Salmon. University of California Press, Berkeley, CA, United States.

Martin, A., Myers, N., Viseu, A., 2015. The politics of care in technoscience. Soc. Stud. Sci. 45 (5), 625−641. https://doi.org/10.1177/0306312715602073.

Mensik, J., Leebov, W., Steinbinder, A., 2019. Survey development: caregivers help define a tool to measure cultures of care. J. Nurs. Adm. 49 (3), 138−142. https://doi.org/10.1097/NNA.0000000000000727.

Minchinton, T.E., Raimondi, P.T., 2001. Long-term monitoring of rocky intertidal communities at the Channel Islands National Park: summary of spatial and temporal trends and statistical power analyses. Sci. Manag. Rev Monitor. Prot. Chan. Islands Natl. Park 22p.

Miner, C.M., Burnaford, J.L., Ambrose, R.F., Antrim, L., Bohlmann, H., Blanchette, C.A., Engle, J.M., Fradkin, S.C., Gaddam, R., Harley, C.D., Miner, B.G., 2018. Large-scale impacts of sea star wasting disease (SSWD) on intertidal sea stars and implications for recovery. PLoS One 13 (3), e0192870. https://doi.org/10.1371/journal.pone.0192870.

Multi-Agency Rocky Intertidal Network (MARINe) Website: Pacificrockyinteridal.org.

Murphy, M., 2015. Unsettling care: troubling transnational itineraries of care in feminist health practices. Soc. Stud. Sci. 45 (5), 717−737. https://doi.org/10.1177/0306312715589136.

Murray, S.N., Ambrose, R.F., Dethier, M.N., 2006. Monitoring Rocky Shores. University of California Press, Berkeley and Los Angeles, California, United States.

Murray, S.N., Weisberg, S.B., Raimondi, P.T., Ambrose, R.F., Bell, C.A., Blanchette, C.A., Burnaford, J.L., Dethier, M.N., Engle, J.M., Foster, M.S., Miner, C.M., 2016. Evaluating ecological states of rocky intertidal communities: a best professional judgment exercise. Ecol. Indicat. 60, 802−814. https://doi.org/10.1016/j.ecolind.2015.08.017.

Overton, W.S., Stehman, S.V., 1996. Desirable design characteristics for long-term monitoring of ecological variables. Environ. Ecol. Stat. 3 (4), 349−361. https://doi.org/10.1007/BF00539371.

Puig de La Bellacasa, M., 2011. Matters of care in technoscience: assembling neglected things. Soc. Stud. Sci. 41 (1), 85−106. https://doi.org/10.1177/0306312710380301.

Puig de la Bellacasa, M., 2012. 'Nothing comes without its world': thinking with care. Socio. Rev. 60 (2), 197−216. https://doi.org/10.1111/j.1467-954X.2012.02070.x.

Puig De La Bellacasa, M., 2015. Making time for soil: technoscientific futurity and the pace of care. Soc. Stud. Sci. 45 (5), 691−716. https://doi.org/10.1177/0306312715599851.

Raimondi, P.T., Wilson, C.M., Ambrose, R.F., Engle, J.M., Minchinton, T.E., 2002. Continued declines of black abalone along the coast of California: are mass mortalities related to El Niño events? Mar. Ecol. Prog. Ser. 242, 143–152. https://doi.org/10.3354/meps242143.

Raimondi, P.T., Orr, D., Bell, C., George, M., Worden, S., Redfield, M., Gaddam, R., Anderson, L., Lohse, D., 2009. Determination of the Extent and Type of Injury to Rocky Intertidal Algae and Animals One Year after the Initial Spill (Cosco Busan): A Report Prepared for OSPR (California Fish and Game).

Raimondi, P.T., Miner, M., Orr, D., Bell, C., George, M., Worden, S., Redfield, M., Gaddam, R., Anderson, L., Lohse, D., 2011. Determination of the Extent and Type of Injury to Rocky Intertidal Algae and Animals during and after the Initial Spill (Dubai Star). A Report Prepared for OSPR. California Department of Fish and Wildlife).

Raimondi, P.T., Ammann, K., Gilbane, L., Whitaker, S., Ostermann-Kelm, S., 2018. Power Analysis of the Intertidal Monitoring Programs at Channel Islands National Park and the Bureau of Ocean Energy Management. Natural Resource Report NPS/MEDN/NRR—2018/1661. National Park Service, Fort Collins, Colorado, United States.

Richards, D.V., Davis, G.E., 1993. Early warnings of modern population collapse in black abalone Haliotis cracherodii, Leach, 1814 at the California Channel Islands. J. Shellfish Res. 12 (2), 189–194.

Schrader, A., 2015. Abyssal intimacies and temporalities of care: how (not) to care about deformed leaf bugs in the aftermath of Chernobyl. Soc. Stud. Sci. 45 (5), 665–690. https://doi.org/10.1177/0306312715603249.

Zee, J.C., 2017. Holding patterns: sand and political time at China's desert shores. Cult. Anthropol. 32 (2), 215–241. https://doi.org/10.14506/ca32.2.06.

CHAPTER 8

Deep-water study partnerships: characterizing and understanding the ecological role of deep corals and chemosynthetic communities in the Gulf of Mexico and northwest Atlantic

Gregory S. Boland
Bureau of Ocean Energy Management (Retired), Sterling, VA, United States

Introduction

This case study represents a series of three collaborations for the investigation of sensitive deep-water ecosystems in the Gulf of Mexico (GoM) and northwest Atlantic initiated by the Bureau of Ocean Energy Management (BOEM). Each was a broad-scale multidisciplinary study involving multiple scientific disciplines necessitating at least some form of partnering among institutions (government, educational or private companies) to assemble the expertise necessary. Detailed descriptions of each partnering effort and evolution will be discussed.

BOEM is the bureau within the Department of the Interior (DOI) tasked through diverse Federal laws with overseeing the management of offshore energy and marine mineral resources and protecting the marine, coastal, and human environments in an economically and environmentally responsible manner. Environmental stewardship is at the core of BOEM's mission including the identification and protection of sensitive biological habitats in US Federal waters. To properly manage these resources, BOEM's Environmental Studies Program (ESP), funded by Congress since 1973, develops, funds and manages rigorous scientific research to help inform policy decisions on Outer Continental Shelf (OCS) energy and mineral resources. BOEM's research mandate is, fundamentally, to assess and understand how the Bureau's decision-making impacts the environment, including the human environment, and how those impacts can be avoided or minimized. The discoveries and results of these studies have been directly utilized to create, or modify, regulatory policies that require avoidance or other mitigations to protect biological habitats, including marine archaeological resources.

The three habitat types that are the subjects of this case study are all in deep ocean environments. Two habitat types in the GoM included deepwater corals and chemosynthetic communities (often overlapping). Research targets in the third study were in the northwest Atlantic and focused on submarine canyons, primarily Norfolk and Baltimore Canyons offshore Virginia and Maryland respectively.

The first habitat category in this case study focused on chemosynthetic communities in the GoM. Chemosynthetic communities are groups of animals living in the deep sea (deeper than 300 m (984 ft)) that live on dissolved gases through a symbiotic association with bacteria living inside their tissues. They are remarkable because these are the only large animals that utilize a food source independent of the photosynthesis that supports almost all other life on earth. The first worldwide discovery of chemosynthetic communities (with megafauna or large animals) was made relatively recently and unexpectedly at hydrothermal vents in the eastern Pacific Ocean during geological explorations of the Galapagos spreading center using the submersible *Alvin* in 1977 (Corlis et al., 1979). Similar communities were first discovered in the Eastern Gulf of Mexico at the bottom of the Florida Escarpment in 1983 (Paul et al., 1984) and later in the Central GoM (Gallaway, 1988; Kennicutt et al., 1985).

It was the responsibility of Minerals Management Service (MMS, prior to becoming BOEM in 2011[a]; BOEM will be used in the following text) to investigate these new and unique communities to enable informed decisions for the management of the oil and gas industry and related potential impacting activities. Initial studies of GoM chemosynthetic communities before the larger partnership study discussed here were funded by BOEM through competitive Federal contracts. Two significant prior studies substantially advanced the understanding of these unique communities through intensive surveys, repeated collections, and experimentation. For convenience, these earlier studies were named Chemo I (MacDonald et al., 1995) and Chemo II (MacDonald, 2002).

The second project in this case study focused on deep-water corals. Deep-water coral habitats have only been investigated intensively in the US since the 1990s. Prior to the subject study, past research had clearly demonstrated that deep-water coral habitats are important biodiversity hotspots and significant biological resources with intrinsic and socio-economic value. In the central GoM, deep-water or cold-water corals are limited to areas of natural hard substrate created by the production of carbonate resulting from bacterial processes at the seabed. These areas constitute less than 1% of the deep GoM seabed. The principal deep-water coral species creating habitats are not dependent on sunlight (no symbiotic photosynthetic algae in their tissues), but complex structures can be formed that also support numerous other species.

The third region in this case study series was located in deep-water submarine canyons of the northwest Atlantic, specifically Norfolk and Baltimore Canyons. Potential offshore oil and gas leasing in the Atlantic was considered in 2008. The most striking feature of the continental margin of the eastern United States is that it is incised by dozens of submarine canyons. Early studies in this vicinity have demonstrated that the submarine canyons along the mid-Atlantic continental slope can include unique biological communities associated with hard substrate exposed in and near the canyon features. Also, evidence of chemosynthetic communities had been reported in one of the subject areas, Baltimore Canyon (Hecker et al., 1980), where they identified canyon walls as unique habitats.

Timing

Conception and development of initial BOEM study profiles, i.e., internal proposals, can occur at any time during a fiscal year, but a draft study profile must be created by the January timeframe for consideration of inclusion in a particular year's Study Development Plan (SDP) and potential progress to contract award the following year. Study profiles approved for inclusion in the annual SDP can become a part of the National Studies List (NSL). Studies that make it to the final NSL are generally intended for funding and contracting, contingent on available funding. Although timing can vary, and some studies can be awarded on an accelerated timeline, it generally takes $1^1/_2$ years from the identification of the need to contract award. In the context of developing a large partnership collaboration, the steps to be taken within this timeframe require planning for when partners need to be cultivated and come onboard for commitment of funding or in-kind support. As partners are identified, they can also play a role in the Statement of Work (SOW) development and proposal and review processes.

Project setup and studies overview

Sequentially, the three partnership studies appearing in Table 10.1 are as follows:

(1) Investigations of Chemosynthetic Communities on the Lower Continental Slope of the Gulf of Mexico (Chemo III);

(2) Exploration and Research of Northern Gulf of Mexico Deepwater Natural and Artificial Hard Bottom Habitats with Emphasis on Coral Communities: Reefs, Rigs, and Wrecks (Lophelia II), and

(3) Exploration and Research of Mid-Atlantic Deepwater Hard Bottom Habitats and Shipwrecks with Emphasis on Canyons and Coral Communities (Atlantic Canyons). All three of these projects involved participation by three Federal agencies, BOEM (as the contracting agency), National Oceanic and Atmospheric Administration's Office of Ocean Exploration and Research (NOAA OER [b]), and the US Geological Survey (USGS).

All three Federal agencies, BOEM, NOAA OER, and USGS had different but equally significant roles in each of the three studies discussed. The evolution of how USGS was involved and became fully integrated into the project goals of these BOEM projects is a significant aspect of this case study (Fig. 8.1).

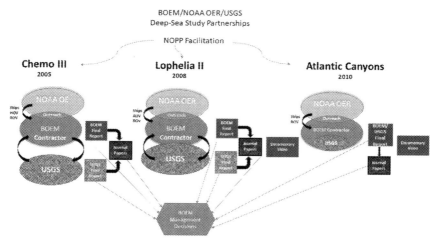

Fig. 8.1 Schematic of project structure and partners.

Partnering agency introductions
NOAA OER
NOAA's Office of Ocean Exploration and Research (OER) through discovery, innovation, and the systematic exploration and scientific study of unknown ocean areas and phenomena serves to ensure that NOAA can meet its goal to "Protect, Restore, and Manage the Use of Coastal and Ocean Resources through an Ecosystem Approach to Management." OER has funded many expeditions in the Gulf of Mexico and Atlantic and seeks innovative scientific objectives that will bring discoveries and allow the public to engage in exploration through education and outreach activities connected to the expedition. This NOAA program is the only designated federal exploration program and partners with other agencies in exploration activities. NOAA OER's internal processes for providing in-kind contributions of vessels, submersibles, or ROVs are complex. Many dedicated staff (often unrecognized) play critical roles in scheduling, planning, and organizing cruises and logistics. NOAA OER Expedition Coordinators are the primary facilitators. Origins of funding for vessels or other facilities are also complex and can include multiple sources within NOAA such as the Office of Marine and Aviation Operations (administers NOAA fleet), or within NOAA OER's budget. In this chapter, all in-kind contributions will be labeled as NOAA OER.

USGS
As one of the agencies of the DOI, USGS is a multi-disciplinary science organization dedicated to the timely, relevant, and impartial study of the landscape, our natural resources, and the natural hazards that threaten us focusing on biology, geography, geology, geospatial information, and water. BOEM coordinates systematically on environmental issues with sister bureaus within DOI and continuously evaluates the potential to match USGS capabilities and interests with the study needs of the BOEM mission.

There is a direct historical connection between an element of the USGS budget and BOEM research funding further enhancing partnering opportunities with BOEM. In 1993, the National Biological Survey (NBS) was created to perform all biology-related research of the DOI. NBS was dissolved in 1995 and was moved into USGS and given a new name, Biological Resources Division (BRD) within the USGS. Limited annual funding dedicated to the mission needs of BOEM (then MMS) was carried forward after the dissolution of NBS and BRD. With the reorganization of USGS and the creation of seven new Mission Areas in 2011, the BRD, and the cyclical budget category identified for BOEM research needs (known as the Outer Continental Shelf (OCS) budget) fell under the new Ecosystems Mission Area. These funds are considered an important part of the planning and mission goals of the BOEM ESP and also provide additional pathways for partnerships between BOEM and USGS.

Each year, the BOEM ESP identifies priority studies that fit within the available USGS OCS budget and bureau expertise that are sent to the Associate Director (AD) of the USGS Ecosystems Mission Area. The process of partnering with USGS scientists in BOEM studies closely follows the internal BOEM study development process and timeline. BOEM scientists that are involved in development of study profiles, potentially leading to contracts, are encouraged to maintain awareness of applicable USGS science capabilities. Ongoing networking through a variety of mechanisms including conferences enhances awareness and stimulates partnership development between BOEM and USGS. This process was followed in all three case study examples. The latter two examples within this case study required relatively detailed knowledge of study components to be taken on by USGS scientists prior to the release of the BOEM contract Request for Proposals (RFP).

Participation of USGS in the BOEM studies described in this case study addresses the overarching Federal mandate for agencies to meet the needs of other agencies as possible. The established association between the bureaus represents a collaborative partnership that uses the expertise of scientists from both agencies to provide peer-reviewed scientific information for BOEM decision-makers.

NOPP

The National Oceanographic Partnership Program (NOPP) is an overarching cornerstone to the partnership success for all three studies presented in this case study by facilitating all three partnerships. NOPP is a collaboration of Federal agencies established by Congress (Public Law 104-201) to provide leadership and coordination of national oceanographic research and education programs. Through facilitated collaborations, Federal agencies can leverage resources to invest in priorities that fall between agency missions or that are too large for any single agency to support. NOPP brings public and private sectors together to support larger, more comprehensive projects, to promote the sharing of resources, and to foster community-wide innovative advances in ocean science, technology, and education.

In addition to the critical fostering of partnerships, NOPP also assisted in the advertising of the availability of the BOEM study solicitations or RFPs of all three projects in addition to standard announcements in FedBizOps (now Contract Opportunities). NOPP also obtained outside scientists for peer review of incoming competitive proposals.

Chemo III

Project implementation

The Chemo III study was conceived by BOEM in early 2004 and was a logical follow-on to the previous two major studies of chemosynthetic communities on the upper slope of the GoM (MacDonald et al., 1995; MacDonald, 2002) and driven by the continuing evolution of offshore technology and capabilities allowing oil and gas exploration and development throughout the full depth range of the GoM. The focus of this study was on chemosynthetic communities on the lower slope of the GoM deeper than 1000 m (3280 ft) and to develop or improve remote assessment methodologies for detecting the presence of the communities and develop some predictive capabilities for avoiding impacts from energy development activities. Other sensitive communities of interest also included deep-water coral habitats in the deepest parts of the GoM where very little was known. Due to the extreme depths involved and facilities required for research to address information needs, funding was a major consideration. Daily costs for a large deep-ocean research vessel combined with an HOV or ROV capable of work below 1000 m (3300 ft) are quite high (up to $80,000 per day).

At the time of early study profile development for the Chemo III study, an extremely fortuitous BOEM partnership association was underway with another study for the first time. In 2003, BOEM began the development of a study of seven historical shipwrecks in the GoM with a very limited budget of approximately $300K. All of the shipwrecks had important significance, many being casualties of World War II (WWII) including the only German U-boat sunk in the GoM, but they were also in relatively deep water (up to 1964 m/6442 ft) requiring expensive vessel and ROV resources. This WWII shipwreck study became BOEM's first significant utilization of NOPP partnership facilitation. At the NOPP table, where member organizations met in Washington D.C., the founding Director of NOAA OE attending the meeting offered significant funds for research relevant to the NOAA OE mission (this is one of the basic functions of NOPP, to promote sharing of resources). The BOEM WWII shipwreck study was a likely fit for a partnership and the Studies Chief (and Chief Scientist) for the BOEM GoM office met with the OE Director to discuss the possibilities. Others in BOEM and NOAA OE were also consulted through numerous conversations. The NOPP and the NOPP Program Director was also supportive. With the limited BOEM funding available, the GoM Regional Director also welcomed outside in-kind financial support. With unanimous support, it was decided that this partnership opportunity would move forward. The incredible characteristic of this major partnership was the fact that there was no written documentation in the form of a Memorandum of Understanding (MOU) or Interagency Agreement (IA) between BOEM and NOAA OE. It was based on the trust established by the long-term working relationship between BOEM and NOAA and historical relationships between all the principal architects of the partnership. NOAA OE funded the research vessel as well as a highly capable ROV that studied six shipwreck sites over 18 days. This highly successful project resulted in the 2007 final report by C&C Technologies, the private firm awarded the BOEM contract (Church et al., 2007).

The great success of this NOPP-facilitated study funded by BOEM and NOAA OE was an inspiration for the challenging Chemo III study. Communication with NOAA OE contacts was initiated at the earliest stage of study development in 2004, the same period of the successful field work for the WWII shipwreck study. This early partnership development was necessary beginning approximately 18 months before the anticipated Chemo III contract award. NOAA OE was very interested and also had available support for the two anticipated field sampling years of Chemo III. Participation of NOAA OE and in-kind funding of research vessels and submergence facilities was critical for the overall success.

One continuing theme of this new powerful partnering relationship between BOEM and NOAA OE, beginning with the WWII shipwreck study, C&C Technologies (Church et al., 2007), is the fact that there was never any written documentation between the funding agencies; no MOUs, no contracts or letters of commitment. It was again based solely on trust as described for the prior WWII shipwreck study. Just one year later, all the principal actors knew each other. With the previous example, the partnering "model" had been successfully demonstrated and this subsequent opportunity was quickly accepted. Beginning with an introduction of the BOEM study need at a NOPP meeting, and a general understanding of NOAA OE's continuing interest in another partnership with BOEM, the process moved quickly. NOAA OE verbally committed to supporting the first year's field expedition of CHEMO III with an understanding for one other field year's support. NOPP brought BOEM and NOAA OER to the table at the right time.

USGS participation with Chemo III was at an early phase in using relatively new USGS deep-water biological discipline capabilities for BOEM needs. BOEM had requested USGS studies of GoM deep-water corals in a previous deep-water coral project (CSA International) including fish trophics, microbiology, and coral genetics. The initial utilization of USGS expertise in Chemo III was for work with deep-sea coral genetics. This collaboration (and the following two partnership examples) is based on a more formal documented study development process than the BOEM/NOAA OE partnering through the identification of USGS capabilities described in the USGS introduction above. A team with the USGS Leetown Science Center had earlier received samples of deep-water corals from the prior WWII shipwreck study (Church et al., 2007) and Lophelia I study (CSA International, 2007), and the lead scientist from that team also participated in the second leg of the Chemo III *Atlantis/Alvin* BOEM/NOAA OE cruise. Other USGS scientists were brought into the Chemo III project in the following year (2007) and teams were funded within USGS for the disciplines of microbiology, benthic and pelagic fish, and benthic macrofauna communities (as illustrated by arrows in Fig. 8.1). This earlier partnering with USGS represented a level of parallel projects with common objectives ("cooperation") compared to fully working together on a common project, i.e., "collaboration" (Interagency Working Group on Ocean Partnerships, 2010).

Contract award

The Bryan, Texas company, TDI Brooks International Inc. was awarded the BOEM contract in August 2005, approximately 18 months after the initial idea for this project was conceived, with subcontractors including Texas A&M University/Corpus Christy, Pennsylvania State University,

Harvard University, Louisiana State University and University of Georgia. The team also included investigators from three other countries, France, Germany and Austria.

Within this study and the following two study examples, an entirely separate universe of partnering evolves within the world of responders to the BOEM RFP in development of their teams and submission of competitive proposals. Expertise in the deep-water science subject areas of these studies is relatively rare. Although details remain confidential, the number of bidding teams is small due to the knowledge pool available. Expertise often comes from outside the United States. Occasionally, some investigators will join multiple bidding teams. Most partnerships within teams is established by subcontract to the prime contractor. Others join the team without salaries only for the additional research opportunity in the deep sea.

The BOEM contract award for Chemo III was for approximately $3.3M. The in-kind contribution value from NOAA OE is difficult to estimate exactly due to variable facility lease rates, but using the cost for a contractor to obtain the same facilities, NOAA OE also provided an in-kind value to the study of approximately $3 million. The genetics team with USGS began collaboration with Chemo III in 2006 with a total budget of $175K. The other subject matter teams including fish trophics, crustaceans, microbiology and paleoecology was later funded within USGS for an additional $473K. USGS also funded the *R/V Cape Hatteras* for $280K for its 2007 cruise, also with the OCS budget for BOEM needs.

The total of all direct and in-kind funding of the project budget came to approximately $7M. Additional undocumented support came from numerous internal sources within all agencies and partner institutions. Total key personnel number at least 25 contributing as authors of the two agency final reports. Many key players in BOEM, NOAA OE, USGS and all the participating institutions were also critical to the success of the project but were not report authors.

Field work included four separate cruises, two of which were provided by NOAA OE including one with the submersible *Alvin* and a second with the ROV *Jason II*. The NOAA OE partnership also included extensive outreach assets as part of field research expeditions. The two major cruises staged by NOAA OE produced companion web pages as "Signature Expeditions" with interface established during the cruises. Access to classroom modules, images and videos remains accessible through NOAA's Ocean Explorer Signature Expedition web pages for each field sampling mission (NOAA Ocean Exploration and Research).

Findings

The ultimate usefulness and fundamental success of this study was to provide the best science to BOEM management for decisions regarding the protection of unique and newly discovered chemosynthetic communities in the GoM. BOEM published three reports and also sponsored a dedicated volume of journal Deep-Sea Research (Roberts, 2010) with 18 papers related to the Chemo III study, including contributions from USGS. The final BOEM report was published in (Brooks et al., 2014), due to extended internal editorial delays. Major topics included remote sensing for the location of habitats, microbiology/biogeochemistry, seep community composition, structure, growth rates and temporal change, meiofauna, and background fauna around seeps. USGS produced its own independent report, although there was significant cooperation and exchange of samples and data as well as the participation of USGS scientists in BOEM/NOAA OE cruises. The separate USGS report resulting from this early BOEM-USGS partnership association was a drawback.

Among the numerous significant results, many species new to science were discovered. Results of this study were directly utilized by BOEM to create or modify existing regulatory policies that require avoidance of a variety of biological habitats including chemosynthetic communities. The results of this study also enhance the understanding of deep-water communities worldwide.

Lessons learned

The biggest lesson learned relates to the highly successful NOPP-facilitated partnership between BOEM and NOAA OE, and joined by USGS. This cooperative facilitation greatly enhanced the achievement of the mission to discover and begin to understand ultra-deep-water biological communities in the GoM not previously known. Trust was by far the most important shared attribute between partners. Potential obstacles, such as major facilities availability and weather problems were not serious. The sustainability of the BOEM science team remained robust throughout the study. All Principal Investigators (PIs) participated as anticipated from the beginning through report production. Most worked under subcontracts to the BOEM prime contractor, but others including international partners working through simple letters of commitment were also dependable.

As with any offshore marine study, success is highly dependent on external, often uncontrollable forces such as weather and equipment failures, there is little that can be done to guarantee the level of success of open-ocean research. The possibility for problems with the staging of ships/HOVs/ROVs was spelled out in the initial solicitation SOW which also included a request to describe possible contingencies. Although contractors would not be held responsible for shortfalls of ships or other major facilities to be provided by NOAA OE, this request was considered important to stimulate analysis. The bottom line: funding agencies would do their best to resolve problems.

The efficiency of utilization of USGS expertise in Chemo III was diminished somewhat due to some lag time at the beginning of Chemo III partnering. Coral genetics had been an ongoing USGS discipline utilized in previous BOEM studies and significant sample and data exchanges were accomplished, but other disciplines were not incorporated until the last year of field sampling in 2007. USGS partnering efficiency was improved in the following case study examples.

Awards

The Chemo III study, also known as the Deep Gulf Habitats Project including BOEM, NOAA OE and USGS was awarded the Cooperative Conservation Award of the Department of the Interior in 2007.

Lophelia II

Project implementation

The Lophelia II study was conceived by BOEM in early 2007 and was a logical follow-on to a prior recently completed major BOEM study of deep-water corals in the GoM (MacDonald, 2002). Of particular significance was determining the probability of where high-density coral communities could be found on exposed hard bottom substrate. While the previous Chemo III study was focused on chemosynthetic communities deeper than 1000 m (3280 ft) with limited work on deep-water corals, this Lophelia II study was primarily intended for work on deep GoM coral habitats (including artificial reefs represented by deep-water energy platforms). An additional aspect of the project included the exploration of shipwreck sites in the deep Gulf with historical and biological objectives.

The successful model of BOEM partnering with NOAA OER and USGS was continued for Lophelia II. The Deputy Director of NOAA OER played a major role in fostering the BOEM-NOAA OER partnership. Again, NOPP was a critical entity to bring BOEM and NOAA OER funding partners together. NOPP was approached for sponsorship of this new partnership after study profile development early in the year concurrent with support of the study through the vetting process and progress onto the BOEM NSL. The study plan was presented at a NOPP meeting mid-year to formally initiate BOEM's desire for partnering. NOAA OER was very interested and the collaboration was established, again founded on trust between the agencies without written agreements, contracts, or letters of commitment.

Lessons learned from the previous effort were applied and for this new project, the interagency partnering with USGS was fully recognized before contract award and established the requirement for the winning Contractor to institute initial and continuing coordination between BOEM contractors and USGS investigators. Topical areas to be considered for performance by USGS expertise were identified. There was an expectation of data and sample sharing between all Contractor PIs and USGS scientists where appropriate.

Although USGS did conduct separate field sampling efforts (and produced a separate USGS final report), there was significant cross-participation between BOEM and USGS investigators (cruise participation, sample, and data exchange) coming closer to the "pure" definition of collaboration: *sharing objectives and working together on a common project* as illustrated by additional overlap and exchange arrows in Fig 8.1. This step in the evolution of USGS participation in a specific BOEM study was far more sophisticated and productive than earlier efforts.

Contract award

The Bryan, Texas company, TDI Brooks International Inc. was awarded the BOEM Lophelia II contract in July 2008 for approximately $3.3M with subcontractors including Pennsylvania State University, Temple University, Woods Hole Oceanographic Institution, Florida State University, and

Louisiana State University. Other private companies included C&C Technologies Inc. and the PAST Foundation. Extensive partnering efforts occur within the private sector as part of the process to create a final team bidding for the BOEM contract. ROV/AUV systems were operated through the University of Connecticut (*Kraken 2*) and Woods Hole Oceanographic Institution (*Jason II* and *Sentry*) provided by NOAA OER. With the inclusion of artificial reefs in this study (deep-water oil and gas structures), seven offshore platform operators were also important partners in the study allowing access to their structures for ROV surveys. The in-kind contribution from NOAA OER is again difficult to estimate in dollars, but using the cost for a contractor to obtain the same facilities, NOAA OER also provided value to the study of approximately $3M. USGS study components generally described within the BOEM/NOAA OER SOW were funded from the USGS OCS budget for the duration of the 4-year study for a total of approximately $4.8M. This included $716K for research vessels and contracting to the University of North Carolina Wilmington and Florida State University.

The total of all direct and in-kind funding of the project budget came to approximately $11.2M. Additional undocumented support came from numerous internal sources within all agencies and partner institutions. Total key personnel numbered at least 36 considering only the authors of the two agency final reports. Many key players in BOEM, NOAA OER, USGS, and all the participating institutions were critical to the success of the project who were not report authors.

A total of five BOEM/NOAA OER sampling cruises were conducted. Three expeditions used both ships and submergence facilities provided by the NOAA OER partnership. USGS and affiliated personnel conducted separate cruises using five research vessels utilizing funding through the USGS OCS budget dedicated to BOEM study needs. As a significant aspect of their partnership, NOAA OER again provided extensive outreach producing Signature Expedition web pages during four of the Lophelia II cruises. These links included images, videos, mission logs, education modules and lesson plans (NOAA Ocean Exploration and Research Expeditions). All of those resources remain available.

Findings

The Lophelia II project resulted in two final reports. One by the BOEM prime contractor (Brooks et al., 2015) and a USGS final report (Demopoulos et al., 2017). This project made significant contributions to the understanding of deep-water corals both in the GoM and world-wide. Some examples of the exceptional diversity of discoveries and new understanding obtained from this research include; (1) new observations at one of the most diverse deep coral habitats in the GoM, also extraordinary in context of ocean acidification chemistry (unanticipated positive calcification rates), (2) improved predictive methodologies for the occurrence of GoM deep-water corals, (3) a 16 m (52 ft) piston core taken at one thriving *Lophelia* mound site demonstrated mound growth over a time period of 300,000 years (Roberts and Kohl, 2018) (4) the deep shipwreck investigation described by one peer-reviewer as the most significant historical deep ship wreck study in the world, and (5) the highly significant discovery of the first impacts of the *DWH* incident on deep-water corals (White et al., 2012).

The platform investigations during the final BOEM/NOAA OER 2012 cruise were also very successful including new records of deep-water coral distribution and growth. The shipwreck component of the study also made highly significant discoveries. These included the confirmation of exceptional coral communities on shipwrecks.

The totality of the study has provided invaluable baseline information in the context of determining future environmental changes from either human-caused or other sources. Excellent scientific design, including repeatable marked stations and photomosaics, were instrumental in allowing the first investigations of potential impacts to deep GoM biological communities from the *DWH* oil spill. Results from this study were responsible for tangible impacts on BOEM's management of natural resources in the Gulf of Mexico. Specifically, a regulatory instrument termed "Notice to Lessees" was modified as a direct application of this partnership-based science to broaden buffer distances from impacting activities to protect deep-water coral communities (NTL, 2009). USGS results were also substantial allowing future comparisons among complex habitats over great depth and geographic ranges. Refined integrated and multidisciplinary sampling design by all partnering teams set new standards in deep-sea exploration and research including extensive protocols for establishing long-term monitoring sites.

In addition to scientific and management goals, a major project component included educational outreach to allow the general public, especially children, to experience and learn about these unique features and ecosystems. Educational outreach was an important aspect and was a required and integrated component of study objectives in the BOEM/NOAA OER SOW. In addition to the NOAA OER Ocean Explorer education modules on cruise Signature Expedition web pages (NOAA Ocean Exploration and Research Expeditions), the contractor's outreach PI team developed a stand-alone package including extensive lesson plans providing teachers with a multi-week program integrating the discoveries and scientific processes from this study (Lophelia II, 2014). A documentary 23-min video available to the public was also produced describing methodologies and many discoveries (Lophelia II, 2013).

Lessons learned

Through the historical and continuing partnerships in previous interagency studies, the overlapping of mutual agency needs was recognized early in the study development process. The partnering assembled in this diverse interdisciplinary project resulted in an unprecedented success of all objectives of the study. Through the leveraging of talent, funding and physical assets, all three of the Federal agencies participating fully recognized the fact that this research could not have succeeded without the partnership.

The partnership was sustained throughout the study by active communication and reaching out between agencies and a shared common passion for the success and integration of so many diverse scientific disciplines. Subcontracts act to hold investigators to their proposed research, but far more relevant is the passion and dedication these scientists have for their disciplines. NOPP was an important entity that facilitated these communications as well as significantly contributing to the contracting processes. There were no funding issues for the duration of the study and in fact, additional funding was provided by BOEM leading to one of the remarkable discoveries of 300,000 years of coral growth at one of the sites (Roberts and Kohl, 2018). NOAA OER provided all facilities to the partnership as anticipated through the trust-based relationship for this study.

This project represented a second step in the integration of USGS partners into a BOEM/NOAA OER deep-water project with significant sharing of not just data and samples but also the substantial participation of USGS scientists on BOEM cruises. Additional complimentary cruises were also conducted by USGS. These separate efforts were not cost-effective and included some other weaknesses as related to the theoretical goals of a single partnership. However, the USGS work did add substantial results to the broad topics of interest and included many important topical subjects not included in the BOEM Contractor's work.

The incorporation of a documentary video production was well received and considered a very positive and constructive choice. Including the requirement for this kind of outreach deliverable should be continued in similar future studies.

Regarding starting over if one could, the preferable partnership structure would be for all investigators to work together on the same study, use the same research vessels, and produce a single final report. This is exactly what was done in the next example in this case study series.

Awards

The Lophelia II study, also known as the Lophelia II: Reefs, Rigs, and Wrecks, including BOEM, NOAA OER, and USGS was awarded the annual NOPP Excellence in Partnering Award in 2012.

Atlantic Canyons

Project implementation

A workshop was sponsored by BOEM in December 2008 to bring experts together to assess the existing scientific knowledge base along the Virginia Coast and the information gaps that need to be addressed should a lease sale for oil and gas activities be held for the Virginia outer continental shelf (Diaz et al., 2009). Submarine canyons were identified as one priority area. This canyon study was conceived by BOEM in early 2009.

Comparable to the previous examples, most of this study was in deep water and required expensive assets for field sampling. The shallowest portions of the canyon features of interest were at 100 m (328 ft) but extended onto the slope at depths of more than 1600 m (5250 ft). The successful BOEM Lophelia II project partnering with NOAA OER and USGS was ongoing as this study was developing. NOAA OER was again considered the ideal partner. NOPP was utilized as the bridging organization to formally bring BOEM and NOAA OER together for another major exploration and research study and solidified through meetings during mid-2009. Planning meetings were held initially with NOAA OER in June 2009 followed by the formal introduction at a NOPP meeting in mid-2009 approximately seven months after conception. The unique trust level between BOEM and NOAA OER continued. Communications and evaluation of USGS interest and capabilities was ongoing with awareness of the prior Lophelia II study winding down.

Partnering with USGS was approached in a new and far more effective way for this Atlantic Canyons study. Through continuing communications with scientists from the appropriate USGS disciplines, a new and untested method for incorporation of USGS into this BOEM/NOAA OER study partnership was established. Work plans outlining four USGS disciplines were included in the BOEM SOW solicitation as attachments for anticipated contributions to be performed by USGS collaborating with the BOEM-funded contractor; (1) benthic ecology, trophodynamics and ecosystem connectivity, (2) genetics studies on deep-water corals and associated communities, (3) paleo-ecology of deep-sea corals, and (4) microbial ecology. These attachments remained a part of the final BOEM contract in addition to a specific SOW Section C Task #3; "Integration of USGS Team." This Task spelled out the direct participation of USGS scientists on cruises, anticipated sample and data sharing and production of a single final report together with the BOEM contractor. Some overlaps of objectives between the contractor's PIs and USGS scientists were anticipated with the expectation that compromises would be have to be made.

This new partnership approach eliminated substantial costs of separate USGS field sampling, insured open communication between all PIs, and eliminated efforts necessary to produce two separate final reports (Fig. 8.1) (in addition to the consolidation of all results into one source for public access). This was an untested model for "embedding" federal scientists with private contractors under a BOEM contract, but successful collaboration was anticipated. Some additional delay in final report preparation was also anticipated due to required internal review processes for publishing by USGS scientists, but the additional time needed was worked into overall schedules as necessary.

Contract award

After development of the initial science needs by BOEM, all three partnering agencies participated in review and finalizing the study's SOW with sponsorship and additional peer-reviews again provided by NOPP. All agencies were also involved in the evaluation of both written and oral competitive proposals and selection of the winning offeror. BOEM was responsible for the contracting and funding of the prime contractor.

The Stuart, Florida company, CSA Ocean Sciences Inc. was awarded the BOEM Atlantic Canyons contract for approximately $3M in September 2010, approximately 18 months after the identification of the study need, with subcontractors including University of North Carolina Wilmington, Florida State University, University of Rhode Island, Texas A&M University, University of Oregon, University of Louisiana at Lafayette and later in the project, student volunteers from Cape Fear Community College. Scientists from two other countries also participated; Bangor University in Whales and the Royal Netherlands Institute of Sea Research. Other entities included the North Carolina Museum of Natural Sciences and the private company Artwork Inc. Extensive partnering efforts occur within the private sector as part of the process to create a final team bidding for the BOEM contract. ROV systems were operated through the University of Connecticut (*Kraken 2*) and Woods Hole Oceanographic Institution (*Jason II*). The estimated in-kind contribution for vessels and these ROVs from NOAA OER was approximately $3M. USGS study components described within the BOEM contract SOW were funded from the USGS OCS budget (designated for BOEM study needs) for the duration of the 5-year study for a total of approximately $3.6M. USGS also provided components of two moorings that were deployed for one year in Baltimore and Norfolk Canyons.

The total of all direct and in-kind funding of the project budget came to approximately $9.6M. Additional undocumented support came from numerous internal sources within all agencies and partner institutions (e.g., two benthic landers provided by Royal Netherlands Institute of Sea Research). Considering only the final report authors (Ross et al., 2017), total key science personnel came to 37. Many key players in BOEM, NOAA OER, USGS and all the participating institutions were critical to the success of the project who were not report authors. A total of 50 key participants were listed in project award nominations.

A total of four sampling cruises were conducted for a total of 90 days at sea, all ship and ROV facilities were provided by NOAA OER. Outreach activities were a part of all cruises with web sites created both by NOAA OER as Ocean Explorer Expeditions as well as the North Carolina Museum of Natural Sciences (one of the subcontractors).

Findings

Results have included the discovery of a large methane seep chemosynthetic community and extensive coral habitats with unexpected densities, and distribution records for the habitat-forming deep-water coral *L. pertusa*. The geochemistry of surface sediment and sediment trap samples supports the hypothesis that the two canyons serve as conduits for transport of shelf sediment and associated organic matter to the deep sea. Significant differences were discovered within and between the two canyons.

Public outreach was an integral part of the Atlantic Deepwater Canyons study from its inception. These efforts consisted of two major components: web-based outreach (through NOAA OER and the North Carolina Museum of Natural Sciences) and the completion of a 24-min high definition video production (Pathways to the Abyss Video, 2015). These components allowed the mission to reach a variety of audiences throughout the entire project. In addition to cruise progress web pages, NOAA OER produced education modules and numerous lesson plans for students in grades 5-12 specifically designed for the project available from the two primary expedition web pages available through NOAA's Ocean Explorer (NOAA Ocean Exploration and Research Expeditions).

Results from this study have been directly utilized in Federal decisions (outside BOEM) to protect canyon habitats including modifications to a National Marine Fisheries Fishery Management Plan and a Presidential Memorandum protecting multiple canyons. Study results also provide the foundation for protective measures considered by BOEM if any energy exploration in this area occurs in the future.

Lessons learned

Although there were some challenges to merge science teams from a government science agency with a private contractor, this partnership methodology was very successful. The trust-based partnership between BOEM and NOAA OER again proved to be exceptional. Production of a single final report was far superior to previous scenarios. Participation and contributions of all scientists greatly enhanced the comprehensive scope of the study and breadth of the science results into a single publication (and dozens of following journal publications) that will remain available to the public.

If starting over, there are not many changes that would be considered. Many complexities were anticipated. Some are unavoidable such as bad weather or ship mechanical issues, but some are more controllable. One inherent conflict was the overlap of subject matter interest and expertise between multiple investigators, potentially between the contractor and USGS PIs. Being the first time this scenario had played out under a BOEM contract, it was a new territory. Everything worked out, but early protocol development (assignment and scheduling of internal USGS reviewers) might have saved some delay in publication.

Although important information resulted, the addition of a major nautical archaeology component to the overall study did significantly reduce asset availability and budgets oriented to other disciplines. The impacts on overall goals should be carefully considered when planning the merging of largely disparate disciplines into a single effort.

Some types of analyses were determined to be inefficient, in one example, video analyses performed by numerous PIs on the same records for narrowly-focused objectives. Pre-planning and collaboration between those needing information from the same video data may have been possible to save considerable effort.

The production of a single, high-quality documentary video was very well received and highly recommended for future large studies, however in hindsight, the addition of one or more separate shorter videos (approximately 5 min) would also fill an important niche for news releases and considering limited time availability for presentations or for many in the general public.

Additional benefits of this new merged paradigm (full collaboration, single report):
- Single contact for overall project communications (Contractor)
- Data exchange more efficient,
- Investigators more comfortable with sharing data,
- Comparisons between PIs made quickly to see how pieces fit together,
- Move to synthesis phase quicker,
- Synthesis greatly improved, and
- Collaboration between all PIs on journal papers simplified.

Awards

The Atlantic Canyons/Pathways to the Abyss study including BOEM, NOAA OER and USGS was awarded both a DOI Partners in Conservation Award in 2013 as well as the annual NOPP Excellence in Partnering Award for 2015.

Summary comments

The tremendous diversity of these three multidisciplinary studies required the expertise from a large number of ocean science sectors. In many large interdisciplinary marine studies, subject areas are often fragmented and not interconnected to components outside of individual investigator's subject areas. These partnerships addressed priority issues that bridge the mandates of individual federal agencies. The fundamental benefit of partnerships is based on the cost benefits of leveraging but even more basic, large issues can be addressed that cannot be accomplished by a single agency or sector.

These partnerships leveraged talent, funding, and physical assets resulting in a far more significant outcome than any individual agency could do on its own. Soliciting competitive proposals enhanced the contribution of cutting edge interdisciplinary science applied to the needs of all agencies. A creative collaboration from a diverse array of scientists and institutions also provided flexibility for supporting new approaches and emerging issues.

Partnerships maximized the utility of research results and extended limited budgets enhancing the ability to achieve conservation goals of all agencies. The specific action that led to these successful partnerships was the active communication and reaching out between agencies with similar conservation and science missions. NOPP was the critical bridging organization for this process.

These three projects unquestionably contributed to the discovery of new species, processes, biological habitats, and culturally significant historical sites in both the GoM and Atlantic. They have been critical to our understanding of how these systems function and in elucidating the potential effects of human activities. In addition to project reports and dozens of scientific journal publications, educational outreach products including lesson plans, education modules and documentary videos will remain available for others to experience and learn about these unique ecosystems.

Acknowledgments

People credited with the success of the three projects in this case study number in the hundreds. Most all individuals, companies and academic institutions have been recognized with awards received by all projects; either the Department of the Interior's (DOI) Conservation Award, the National Oceanographic Partnership Program's Excellence in Partnering Award, or both. Particularly deserving of individual mention are the Contracting Officers: Debra Bridge, Mary Coleman and Christy Tardiff. Key players in NOAA included the forward thinking and leadership of John McDonough for all three projects, and also the essential engagement of project coordinators for NOAA: Jeremy Potter, Felipe Arzayus, John Tomczuk, Emily McDonald, and Kasey Cantwell. Within DOI, William Shedd and Jesse Hunt with the Resource Evaluation Section of Minerals Management Service played a critical role in providing locations for likely deep-water coral and chemosynthetic community locations in the Gulf of Mexico using in-house industry seismic data. Special thanks also go to Bureau of Ocean Energy Management managers that made these studies possible, especially James Kendall and Rodney Cluck.

References

Brooks, J.M., Fisher, C., Roberts, H., Bernard, B., McDonald, I., Carney, R., Joye, S., Cordes, E., Wolff, G., Goehring, E., 2014. Investigations of chemosynthetic communities on the lower continental slope of the Gulf of Mexico: Volume I: Final report. U.S. Dept. of the Interior, Bureau of Ocean Energy Management, Gulf of Mexico OCS Region, New Orleans, LA. OCS Study BOEM 2014-650. p. 560. Available from: https://espis.boem.gov/final%20reports/5406.pdf.

Brooks, J.M., Fisher, C., Roberts, H., Cordes, E., Baums, I., Bernard, B., Church, R., Etnoyer, P., German, C., Goehring, E., McDonald, I., Roberts, H., Shank, T., Warren, D., Welsh, S., Wolff, G., Weaver, D., 2015. Exploration and research of northern Gulf of Mexico deepwater natural and artificial hard-bottom habitats with emphasis on coral communities: reefs, rigs, and wrecks— "Lophelia II" final report. In: U.S. Dept. of the Interior, Bureau of Ocean Energy Management, Gulf of Mexico OCS Region, New Orleans, LA. OCS Study BOEM 2016-021, p. 628.

Church, R., Warren, D., Cullimore, R., Johnston, L., Schroeder, W., Patterson, W., Shirley, T., Kilgour, M., Morris, N., Moore, J., 2007. Archaeological and biological analysis of world war II shipwrecks in the Gulf of Mexico: artificial reef effect in deep water. In: US Dept. of the Interior, Minerals Management Service, Gulf of Mexico OCS region, new Orleans, LA. OCS Study MMS, 15, p. 387.

Corliss, J.B., Dymond, J., Gordon, L.I., Edmond, J.M., von Herzen, R.P., Ballard, R.D., Green, K., Williams, D., Bainbridge, A., Crane, K., van Andel, T.H., 1979. Submarine thermal springs on the Galapagos rift. Science 203 (4385), 1073–1083.

CSA International, Inc., 2007. Characterization of Northern Gulf of Mexico Deepwater Hard Bottom Communities with Emphasis on Lophelia Coral. U.S. Department of the Interior, Minerals Management Service, Gulf of Mexico OCS Region, New Orleans, LA. OCS Study MMS 2007-044:169 pp. + app. Available from: https://espis.boem.gov/final%20reports/4264.pdf.

Demopoulos, A.W., Ross, S.W., Kellogg, C.A., Morrison, C.L., Nizinski, M.S., Prouty, N.G., Borque, J.R., Galkiewicz, J.P., Gray, M.A., Springmann, M.J., Coykendall, D.K., 2017. Deepwater Program: Lophelia II, continuing ecological research on deep-sea corals and deep-reef habitats in the Gulf of Mexico. In: US Department of the Interior, Geological Survey.

Diaz, R.J., Brill, R., Schaffner, L.C., Able, K.W., Atkinson, L., Austin, D., Kraus, S., Lipton, D., 2009. Workshop on Environmental Research Needs in Support of Potential Virginia Offshore Oil and Gas Activities.

Gallaway, B.J., 1988. Northern Gulf of Mexico Continental Slope Study. Minerals Management Service, Final Report. Year 4, Volume II: Synthesis Report. OCS Study/MMS 88-0053.

Hecker, B., Blechschmidt, G., Gibson, P., 1980. Epifaunal zonation and community structure in three Mid-and North Atlantic canyons. In: Final report for the Bureau of Land Management. United States Department of the Interior. Contract Opportunities: Official U.S. government website for people who make, receive, and manage federal awards. Available from: https://beta.sam.gov/ (Accessed May 1st, 2021).

Interagency Working Group on Ocean Partnerships, 2010. Strategic Plan. Available from: https://www.nopp.org/wp-content/uploads/2010/03/IWG-OP-Strategic-Plan1.pdf.

Kennicutt, M.C., Brooks, J.M., Bidigare, R.R., Fay, R.R., Wade, T.L., McDonald, T.J., 1985. Vent-type taxa in a hydrocarbon seep region on the Louisiana slope. Nature 317 (6035), 351–353.

Lophelia II, 2013. Video Production Resulting from Partnership Study, Exploration and Research of Northern Gulf of Mexico Deepwater Natural and Artificial Hard Bottom Habitats with Emphasis on Coral Communities: Reefs, Rigs, and Wrecks.

Lophelia II, 2014. Understanding Deep Sea Coral DVD: curriculum, challenge scenario maps, photo mosaics, taxonomic keys. In: Produced as part of Bureau of Ocean Energy study, Exploration and Research of Northern Gulf of Mexico Deepwater Natural and Artificial Hard Bottom Habitats with Emphasis on Coral Communities: Reefs, Rigs, and Wrecks with Additional Funding From NOAA OER and Collaboration With USGS.

MacDonald, I.R. (Ed.), 2002. Stability and Change in Gulf of Mexico Chemosynthetic Communities. Volume II: Technical Report. Prepared by the Geochemical and Environmental Research Group, Texas A&M University. U.S. Dept. of the Interior, Minerals Mgmt. Service, Gulf of Mexico OCS Region, New Orleans, LA. OCS Study MMS 2002-036:456 pp. Available from: https://espis.boem.gov/final%20reports/3072.pdf.

MacDonald, I.R., Schroeder, W.W., Brooks, J.M. (Eds.), 1995. Chemosynthetic Ecosystems Studies Final Report. Prepared by Geochemical and Environmental Research Group. U.S. Dept. of the Interior, Minerals Management Service, Gulf of Mexico OCS Region, New Orleans, LA. OCS Study MMS 95-0023:338 pp. Available from: https://espis.boem.gov/final%20reports/3323.pdf.

NOAA Ocean Exploration and Research Expeditions. Available from: https://oceanexplorer.noaa.gov/explorations/explorations-by-year.html.

NTL, 2009. Notice to Lessees and Operators of Federal Oil, Gas, and Sulphur Leases and Pipeline Right-of-Way Holders, Outer Continental Shelf, Gulf of Mexico OCS Region: Deepwater Benthic Communities. NTL 2009-G40.

Pathways to the Abyss Video, 2015. Video Production Showcasing Two Deep-Water Atlantic Canyons from the Study, Exploration and Research of Mid-Atlantic Deepwater Had Bottom Habitats and Shipwrecks with Emphasis on Canyons and Coral Communities.

Paull, C.K., Hecker, B., Paull, C.K., Hecker, B., Commeau, R., Freeman-Lynde, R.P., Neumann, C., Corso, W.P., Golubic, S., Hook, J.E., Sikes, E., Curray, J., 1984. Biological communities at the Florida Escarpment resemble hydrothermal vent taxa. Science 226 (4677), 965—967.

Gulf of Mexico cold seeps: deep Sea research Part II. In: Roberts, H.H. (Ed.), Top. Stud. Oceanogr. 57 (21—23), 1835—2060.

Roberts, H.H., Kohl, B., 2018. Temperature control of cold-water coral (Lophelia) mound growth by climate-cycle forcing, Northeast Gulf of Mexico. Deep Sea Res. Oceanogr. Res. Pap. 140, 142—158.

Ross, S., Brooke, S., Baird, E., Coykendall, E., Davies, A., Demopoulos, A., France, S., Kellogg, C., Mather, R., Mienis, F., Morrison, C., Prouty, N., Roark, B., Robertson, C., 2017. Exploration and research of mid-Atlantic deepwater hard bottom habitats and shipwrecks with emphasis on canyons and coral communities: Atlantic deepwater canyons study. Final Report. In: U.S. Dept. of the Interior, Bureau of Ocean Energy Management. OCS Study BOEM 2017-060, p. 1000p.

White, H.K., Hsing, P.Y., Cho, W., Shank, T.M., Cordes, E.E., Quattrini, A.M., Nelson, R.K., Camilli, R., Demopoulos, A.W., German, C.R., Brooks, J.M., 2012. Impact of the Deepwater Horizon oil spill on a deep-water coral community in the Gulf of Mexico. Proc. Natl. Acad. Sci. U.S.A. 109 (50), 20303—20308.

CHAPTER 9

Adaptation to repetitive flooding: expanding inventories of possibility through the co-production of knowledge

Elizabeth K. Marino[a], Annie Weyiouanna[b] and Julie Raymond-Yakoubian[c]

[a]Oregon State University - Cascades, Bend, OR, United States; [b]Bering Strait School District, Shishmaref, AK, United States; [c]Kawerak, Inc., Nome, AK, United States

Introduction/background

Coastlines throughout the world are highly productive places which hold and support many diverse human communities, and account for the homelands of over one billion people (Hauer et al., 2016; Neumann et al., 2015). Because of their large populations and vulnerability to flooding under conditions of climate change (Hauer et al., 2016), understanding adaptation to flooding conditions is a critical part of our collective response to sea level rise. When flooding becomes a normative condition—an event that occurs with regularity and is not an aberration—then these events are collectively understood as repetitive flooding. Current population-based estimates predict that between 4.2 and 13.1 million people in the United States alone are at risk of water inundation given current projections of sea level rise and repetitive flooding (Hauer et al., 2016).

Despite significant predictions that relocation away from coastal areas may be necessary in the future, there are relatively few policy mechanisms available to proactively relocate people away from coastal areas; and researchers have suggested that there is currently no state or federal agency, or regulatory governance framework, to relocate communities *as communities* instead of as individual property owners (Bronen and Chapin, 2013; Marino 2018; Maldonado and Peterson). This is of particular concern to Indigenous communities who have experienced the traumas of forced relocation in their histories (Walters et al., 2011); and to social researchers who have noted that displacement can frequently result in negative social consequences for individuals and communities across generations (Cernea, 1997). Additionally, research has demonstrated that when group-relocation schemes are enacted, but do not take culturally embedded lives and livelihoods into account—such as with some relocation attempts in the aftermath of the 2005 tsunami (CADAN; 2017)—people will eventually abandon newly established settlements because the geographical locations of new village sites, or the organization of infrastructure, does not allow people to reform the socioeconomic and socioecological relationships upon which their communities are based.

These conditions, taken together, suggest that there is an urgent need to understand what adaptation possibilities there are in response to sea level rise; and whether and how relocation will be an option that is adaptive for individuals and communities. In the social science literature, adaptation is sometimes distinguished from coping (Marino and Lazrus, 2015); the former being a structural adjustment in people's lives which leads to a quality of life self-reported to be both coherent and satisfactory, and the latter indicating a structural adjustment that leaves individuals alive, but dissatisfied and with a greater experience of incoherence, which is an indicator of trauma. When communities and social scientists ask questions about adaptation to sea level rise, we are asking about what is possible, what we might call "inventories of possibility," for people facing extreme flooding, so that communities may select those possibilities which will afford them a reasonable, recognizable future.

Defining adaptation in these terms, we argue, means that it would be impossible to understand adaptation to climate change without co-producing knowledge. Trying to establish what communities should do, what site might be a reasonable one to expand or move infrastructure to, or what stages of relocation might be appropriate cannot be established from the "outside." While outsiders may have ideas about how to "cope" with high water and flooding, what would be "adaptive" to community members is, by definition, only approachable by understanding internal, culturally mediated solutions to risk and demands on a reasonable future.

In Alaska, repetitive flooding is not a prediction of a distant future, but a regularly occurring event for many rural communities. Our work is situated within legal structures which outline responses to flooding; and within the community of Shishmaref, Alaska. First, our research aims to create a policy map for adaptation support, including how disaster policies and disaster funding in the United States delineate what is possible to accomplish under current regulatory structures. Second, we are outlining adaptation possibilities as seen by residents in Shishmaref, and what is locally considered a viable and sustainable future for the community under conditions of repetitive flooding. In the first data set, we identify what is possible from a policy perspective. In the second, we identify what is possible from a community standpoint of the culturally mitigated good. By comparing the two we can see where there is overlap, indicating possible paths forward without regulatory change. Where there is a lack of overlap, we hypothesize, will indicate where policies are ineffectual for Inupiat communities and, possibly, where policies are articulating western ethnocentric beliefs that run counter to Inupiat adaptation structures. As a function of the law, "western" here indicates ideas that are primarily rooted in British traditions and British common law that became part of codified legal structures of the United States, including ethnocentric assumptions about the interactions between citizens and governments.

In this project, co-production functions as a way to address the complex problem of adapting to repetitive flooding in a community in western Alaska. Co-production gives insight into why policies that attempt to address repetitive flooding might fail some communities and not others (Marino, 2018). In our case, co-produced research recognizes how different vantage points of knowing (law, scientific knowledge[1], Inupiat knowledge sets) offer the potential for identifying unique adaptation possibilities. We hypothesize that co-producing this research will elicit expanded "inventories of possibility" than would be impossible to catalog either with a policy-only analysis or a community-only analysis.

Throughout this chapter we will outline the stages of collaboration we have participated in to better understand the dynamics of co-producing a set of research questions and outcomes.

---

[1] The authors of this chapter and the editors of this book realize that there are multiple ways of engaging and naming different knowledge sets (science, Indigenous knowledge) for purposes of comparison; and that knowledge sets may overlap in some assumptions, methodologies, and practices. By science we imply the pursuit of new understanding and knowledge that understands itself to be, and is broadly recognized as, following the scientific method. By Indigenous Knowledge we refer to "Indigenous peoples' systems of observing, monitoring, researching, recording, communicating, and learning and their social adaptive capacity to adjust to or prepare for changes" (Jantarasami et al., 2018). We recognize that other authors and communities might refer to these two knowledge systems using other expressions and definitions.

Project setup and goals

Developing the idea

Through the course of our careers, and as participants in collective aspirations toward the co-production of knowledge in the Arctic, we've observed that a primary obstacle for truly co-produced research is how to collaborate in the idea development stage of a research project. When communities or local collaborators are physically distant, technologically disparate, and/or not sharing knowledge or social platforms, it becomes a challenge to form collaborative teams to share ideas in the beginning of a research project. This problem is particularly salient when collaborators are co-constructing a grant proposal (Finders et al., 2016 for an overview of risks).

Our collaborative proposal was written after years of informal discussion—on social media platforms, informal professional settings, and on the phone. The genesis was the collective recognition of the ongoing risks that repetitive flooding posed to Shishmaref, and the inadequacy of policy options to respond well to Indigenous communities under these circumstances (Marino, 2018). Our research questions are about the inherent assumptions embedded in culturally mediated legal possibilities, and the culturally mediated understanding of what is a good future for Shishmaref residents; but the ability to have these conversations was a product of long-term relationship-building among the authors on this chapter, and the Co-PIs on our larger research effort. Ms. Weyiouanna, Dr. Marino, and Dr. Raymond-Yakoubian have known one another and had established trust before our current collaboration. We've met one another's kids and parents. We've exchanged gifts. These informal relationship ties made the process of formal collaboration easier.

Despite these relationships, the timeline for creating our research grant proposal still took longer than we anticipated. One particularly useful tool for ensuring we met the standards for co-production (full participation of all collaborators from the beginning), and did not rush collaborative planning stages, was that the Arctic Social Science Program (ASSP), under the Office of Polar Programs (OPP) at the National Science Foundation (NSF), has a rolling deadline. This seemingly insignificant feature of the ASSP was, we found, foundational to the goals of co-production. What we intended to be an October submission became a February submission. The process of collaborating to identify research questions, getting formal permission from the tribe and other political entities in Shishmaref, getting formal permission from the regional Indigenous nonprofit corporation, and a partner, Kawerak, and ensuring that all of our collaborators had reviewed and commented on the multiple stages of the application, meant a delayed submission date. Because the deadline for submission was not firm, it allowed us all to be patient. In the end we received a three-year NSF grant that included one Co-PI from Shishmaref, two academic Co-PIs, and two Co-PIs from Kawerak, the regional Indigenous Non-Profit Corporation. We are a team of two anthropologists, one local knowledge holder, an attorney, and a public policy specialist. Included in our grant are funds to bring 8–10 Shishmaref community leaders together to vet the materials Co-PIs collected to ensure it captures the intent of the community.

Approval and consent

Many tribes in the Bering Strait region do not have a tribe-specific procedural map for researchers to get permission to do research in traditional Indigenous territory, to begin a collaborative research process, or to approach a tribe about an idea for research. However, getting permission for being in a community, permission for being on tribal territory, and information sharing about how data will be collected and distributed, is possible within the structure of the tribe, a designated political body under the Indian Reorganization Act (IRA). In Alaska, tribal organizations are known as IRAs.

For our collaborative project, Ms. Weyiouanna acted as a liaison between our team and the Shishmaref IRA. We discussed the project, the research objectives, the methodologies, and the timeline, and she brought the project for a formal vote of approval by the IRA. In the letter of approval, the tribe acknowledged both that they approved the research and that they approved Ms. Weyiouanna to act as the formal liaison between the research team and the tribe.

One nuance for thinking through co-production of knowledge is that having a collaborative partner or partner institution who is co-designing the research is distinct from eliciting formal institutional approval for research from local authorities. Both are important, but different, aspects of a project. It would be overly burdensome to place the ethical obligations or success of an entire research team on the back and on the name, of one community member, or even one organization. In our case, while Ms. Weyiouanna is our co-PI and Dr. Raymond-Yakoubian is part of Kawerak, our research project still sought formal approval from the Shishmaref IRA, the Kawerak Board, and other governing entities. This approval allowed our local and regional Co-PIs to share responsibility for the outcomes of the project.

Budget allocation

In our research project, all principal investigators and co-principal investigators are given similar salary allocations per year of the project. This is true regardless of education level or project role. It is our position that the co-production of knowledge cannot be sincere without the recognition that the skill sets and knowledge sets that collaborators bring to a project have a similar economic value.

Budget allocations in our research project are constructed on three tiers. The first is the salary for all co-investigators, as previously mentioned. The second is honorariums for interviewees, for whose time we compensate. The third is a consultation with tribal leaders in two, two-to-three-day overview meetings, to review the data we've collected and analyzed to ensure it meets their standards and to consult on what is possible to make public.

In putting together this chapter, Dr. Marino and Ms. Weyiouanna spoke about compensation during research projects and Ms. Weyiouanna laughed and said, "don't be cheap!" In recent conversations with other Indigenous scientists, the phrase that came up, again and again, is that the time of unpaid volunteerism is over. Put another way, the process of collaboration or consultation with tribes has often involved volunteering on the part of the tribe with little or no compensation from the researcher (more about this in lessons learned). Researchers who want to collaborate have been cheap—treating their collaborator's knowledge as a free resource. The lack of value is one mechanism for rendering some pieces of knowledge legitimate and others illegitimate. We are not arguing here that economic value is a proxy for ultimate value. The value of all knowledge is not equivalent to its going rate on the marketplace (Caffentzis 2008), but collaborative projects should challenge the hierarchal structure that pays scientists and not Indigenous knowledge holders.

Project implementation

Co-production under conditions of uncertainty

Our research project began on September 1, 2019, and was initially dependent on in-person collaborative meetings between constitutional scholars, policy-makers, seal hunters, Inupiat knowledge holders, local leaders, regional leaders, and more. Our plan was to have simultaneous data collection mechanisms occur through (1) consultation with legal experts, trained in particular areas of the law; (2) textual analysis on policy and legal documents; and (3) interviews with Shishmaref residents about their visions for the future. Due to the COVID-19 pandemic, some of our methodological structure has been altered. While more complicated, the textual and legal analysis of our project has remained largely unchanged.

Central to our methodological strategy, however, was what we called a "decolonized" interview session. Because the question/answer dynamic is not a culturally universal mechanism for information exchange; we proposed to have our Co-PI, Annie Weyiouanna, discuss the topics we were interested in (visions for the future, culturally informed adaptation to repetitive flooding), in conversational style and locations which were culturally appropriate to community members in Shishmaref. These "interviews" were to be recorded and would have the same consent process as a more traditional data collection format (in part because we couldn't find a better way to do this). In collaboration with the IRA, we decided that any "visiting" for the purposes of research was not appropriate under pandemic conditions. Even for people Ms. Weyiouanna would otherwise see, we did not want our research project to be responsible for any spread of the virus.

This awareness on the part of all researchers to resist pushing the project forward came in part, again, from Co-PIs long-term investment, or life in, the community. We are aware of the history of pandemics on the Seward Peninsula (Ganley, 1989). In 1918, over half of residents in some nearby Indigenous communities died as a result of the 1918 Flu. Pandemic experiences are an active memory of elders and are part of local oral histories. Historical awareness is not to be taken lightly in co-produced projects. We recognize that the Co-PIs have different lived experience and emerge from different histories. Knowing the history of the region was particularly salient in this case, but we feel is always important. Especially for projects that happen in rural landscapes with Indigenous peoples, it is incumbent upon outside researchers to understand the history of place.

Changing methods mid-stream

Because we have stalled on our decolonized interview sessions, we have added an additional data collection tool. We have begun to work with a local Inupiat photographer and videographer in Shishmaref. We are currently going through his collection of videos and photographs, taken over a 20-year period, and having ongoing discussions with him about his selection of subjects. Included in these images and videos are real-time videos of storms, shots of erosion over time, and Inupiat traditions that are central to daily life in Shishmaref. We have paired this artist with an Indigenous artist and museum curator from a tribe in Louisiana, and they continue to have ongoing discussions about the differences in their vision of climate change, tradition, and the future compared with journalists who have documented the stories of climate change in their communities. We are collaborating now on a museum exhibit of these different visions.

There are multiple purposes behind this new data collection strategy. First, by contrasting journalists' stories with local artists' and residents' stories, we can see a distinction in the normative assumptions of the two groups. Second, once meetings and interview sessions are able to begin again, we will use the photographs as a point of entry into conversation, which is a proven methodology (Wang and Burris, 1997). There is, however, a third, critical reason for this new collaboration. The inclusion of artists as producers of knowledge helps to prevent what we are frequently warned against from our Indigenous colleagues—that taking stories out of a community without adequate approval is the theft of tradition and culture.

Author Raymond-Yakoubian and her colleagues have written a critical piece on co-production that is forthcoming (Yua et al. in progress) that includes important ideas about collecting, storing, and sharing information. The nuances of these questions, and the ongoing battle for intellectual property rights of Indigenous knowledge holders (Mauro and Hardison, 2000), are too complex for us to approach in the limited space of this chapter; however, the presence of local artists and Co-PIs as knowledge holders raises the specter of intellectual property rights and intellectual theft in ways that are salient to our academic colleagues.

The ability to be flexible with the imposition of COVID-19, and to add the work of a new collaborator and artist, functionally occurred because of two primary factors. First, the Lead PI, Elizabeth Marino, has spent time in the community of Shishmaref and can approach and speak with friends there about possibilities. The second is because the artist and Lead PI are friends on social media. Collaboration over social media has been critical to many parts of this project; and allows for quick, real-time communication and back and forth engagement. The lead PI and our collaborators in Shishmaref are in contact on average twice a week via social media. Questions, data, relationship building, and frustrations are all shared in this way.

Ongoing lessons

We are in the early stages of our project; and so final outcomes—both in terms of scientific production and our ability to meet our own standards of ethics—are still uncertain. Despite our infancy, we have already encountered challenges to co-production, which we will outline here.

First, it has been a challenge to create research products that are equitably produced and distributed without undue burden or alienation on the part of one or more of the collaborators. Case in point, this chapter is being written in a hurry. Consultation with the author team was hurried. If time is a culturally mediated phenomenon (Marino and Lazrus, 2016), then the academic author on this piece put demands on the nonacademic authors of this piece that were more consistent with academic lifeways and timeframes than nonacademic lifeways and timeframes. This fault is doubly-problematic because academic writing is not made for a nonacademic audience. There will be academic and nonacademic products as a result of our project—but there is a gap in our understanding about what it means to "co-produce" written products that do not have substantive meaning, or social reward, within the social lives of non-academic colleagues and co-authors.

The second challenge was with the Institutional Review Board (IRB), an academic institution which oversees research that involves human subjects to ensure no unreasonable amount of harm is incurred. Much to our surprise, we were in negotiations with the IRB for over a year, and IRB reviewers were particularly concerned that inclusion of tribal members as researchers, located within a small community, would put undue pressures on research participants. We disagreed with the IRB and experienced this impasse as a complex disagreement over a normative assumption embedded within the research imagination: of objectivity and data collection uncontaminated by personal relationship or power dynamics. We questioned the legitimacy of this concern; and the assumption that power dynamics between researcher and subject were less significant when that researcher was an unknown, unaffiliated person or institution. However, to be approved we had to include language in all of our consent forms that when the researcher was part of a tribal organization, that personal relationships would not be impacted by the outcomes of this research, and that there would be no additional punishment or reward as an outcome of this research. We wonder what it would mean for nonlocal researchers to meet those same standards! Going forward it would be useful to catalog interactions between co-production research teams and their IRB interactions.

One other lesson we are grappling with is how value is allocated in a research project. In standard social science research, and in the way our project is organized, the people collecting and analyzing information are most central to the project and receive salary for their work. In recent weeks this has been questioned by some of our Inupiat collaborators, pointing out that people who hold the stories, not necessarily people who collect stories, are where work exists and that value should be allocated as such. It has been suggested to us that a lifetime of knowledge and experience goes into the moment at which someone may express their vision of the future and offer insight into the ecological changes on the landscape. That we are paying these participants honorariums and the listeners salary is something to consider.

Conclusion: a more collaborative science

Bringing diverse knowledge sets into conversation with one another has much promise. If scientists are interested in innovation, then the moments in which one set of epistemological limits are reached, and then breached, by the epistemological limits of another are critically important. A central thesis in anthropology is the idea that a person is blind to their own bounded assumptions until they encounter the assumptions of another. In the case of our project, the limitations of US federal policy to respond to climate change outcomes become illuminated when they sit alongside Inupiat responses to climate change. Putting these knowledge sets side by side is useful in understanding how adaptation possibilities are limited by our own imaginations.

Acknowledgments

This project is sponsored by NSF grant award # 1921045. The authors would like to thank the many friends and family members who have supported this work. We would also like to thank the Shishmaref Elder's Council, the Shishmaref IRA, the Shishmaref Native Corporation, and the City of Shishmaref.

References

Bronen, R., Chapin, F.S., 2013. Adaptive governance and institutional strategies for climate-induced community relocations in Alaska. Proc. Natl. Acad. Sci. U.S.A. 110 (23), 9320–9325.

Caffentzis, G., 2008. A Critique of Commodified Education and Knowledge (From Africa to Maine). Russell Scholar Lecture, University of Southern Maine, Portland, ME, February, 12.

Cernea, M.M., 1996. The risks and reconstruction model for resettling displaced populations. MM Cernea.
Culture and Disaster Action Network (CADAN), Marino, E., Koons, A., Olson, L., Browne, K.E., Faas, A.J., Maldonado, J., 2017. A Helping Hand. The Mark News.
Flinders, M., Wood, M., Cunningham, M., 2016. The politics of co-production: risks, limits and pollution. Evid. Policy A J. Res. Debate Pract. 12 (2), 261–279.
Ganley, M.L., 1998. The dispersal of the 1918 influenza virus on the Seward Peninsula, Alaska: an ethnohistoric reconstruction. Int. J. Circumpolar Health 57, 247–251.
Hauer, M.E., Evans, J.M., Mishra, D.R., 2016. Millions projected to be at risk from sea-level rise in the continental United States. Nat. Clim. Change 6 (7), 691–695.
Jantarasami, L.C., Novak, R., Delgado, R., Marino, E., McNeeley, S., Narducci, C., Whyte, K.P., 2018. Tribes and Indigenous Peoples. In: Reidmiller, D.R., Avery, C.W., Easterling, D.R., Kunkel, K.E., Lewis, K.L.M., Maycock, T.K., Stewart, B.C. (Eds.), Impacts, risks, and adaptation in the United States: Fourth National Climate Assessment, volume II. U.S. Global Change Research Program, Washington, DC, pp. 572–603.
Marino, E., 2018. Adaptation privilege and voluntary buyouts: perspectives on ethnocentrism in sea level rise relocation and retreat policies in the US. Global Environ. Change 49, 10–13.
Marino, E., Lazrus, H., 2015. Migration or forced displacement? The complex choices of climate change and disaster migrants in Shishmaref, Alaska and Nanumea, Tuvalu. Hum. Organ. 74 (4), 341–350.
Marino, E., Lazrus, H., 2016. We are always getting ready": how diverse notions of time and flexibility build adaptive capacity in Alaska and Tuvalu. Contextualizing Disaster, Berghahn, New York.
Mauro, F., Hardison, P.D., 2000. Traditional knowledge of indigenous and local communities: international debate and policy initiatives. Ecol. Appl. 10 (5), 1263–1269.
Neumann, B., Vafeidis, A.T., Zimmermann, J., Nicholls, R.J., 2015. Future coastal population growth and exposure to sea-level rise and coastal flooding-a global assessment. PloS one 10 (3), e0118571.
Walters, K.L., Mohammed, S.A., Evans-Campbell, T., Beltrán, R.E., Chae, D.H., Duran, B., 2011. Bodies don't just tell stories, they tell histories: embodiment of historical trauma among American Indians and Alaska natives1. Du Bois Review: Social Science Research on Race 8 (1), 179–189.
Wang, C., Burris, M.A., 1997. Photovoice: concept, methodology, and use for participatory needs assessment. Health Educ. Behav. 24 (3), 369–387.

CHAPTER 10

Lessons learned from nine partnerships in marine research

Francis K. Wiese[a], Guillermo Auad[b], Elizabeth K. Marino[c] and Melbourne G. Briscoe[d]

[a]Stantec Consulting Services, Inc., Anchorage, AK, United States; [b]Office of Policy and Analysis, Bureau of Safety and Environmental Enforcement, U.S. Department of the Interior, Sterling, VA, United States; [c]Oregon State University - Cascades, Bend, OR, United States; [d]OceanGeeks LLC, Alexandria, VA, United States

Introduction

Nine marine research partnerships different in scope, geographical focus, duration, partner diversity, and budgets are examined in the previous chapters. All of them emerged because of the collective will of researchers and managers at multiple institutions, trying to achieve goals that ultimately would not have been achievable by any one person or institution working alone. Motivations to initiate or agree to these partnerships varied, and included exploring new and broader needs or questions, avoiding redundancies, leveraging resources, and identifying joint or complementary goals. While many of these partnerships attracted new partners by asking two simple questions, *"what do you need?"* and *"what can you contribute?"*, in other cases this understanding was implicit and the engagement was more spontaneous. An important common aspect of all these partnerships was that all partners contributed resources and received outcomes from what was collectively achieved.

In this chapter, we aim to use the nine case studies to define elements of successful and sustainable partnerships and extract lessons learned by asking what worked, what didn't work, what barriers existed to create and maintain partnerships, how did partners overcome these obstacles, or which ones could not?

In this context, it is important to highlight the triple social process that is engaged when partnerships focus on scientific research. On one side, the act of partnering is a social process targeting common goals and distributing multiple benefits for all involved in a collaborative manner. But when partnering involves a scientific enterprise, the social process of science itself is triggered and its centuries-old social norms of universalism, communality, disinterestedness, and organized skepticism are automatically invoked (Merton, 1973). Finally, results and findings during and after a given partnership may impact society, thus triggering a process of broader knowledge sharing and integration into societal processes. This sharing and integration could take several forms, including the use of scientific research products for policy- and decision-making; increasing community resilience to environmental and anthropogenic pressures; sustainable use practices of ocean resources, or any other aspects, originally planned or not, that result in societal benefits. Given that partnerships targeting scientific research combine more than one social process, the early involvement of social scientists in these marine partnership programs positively impacts their creation, development, sustainability, and application.

Analysis

The main programmatic elements of the nine case studies presented in the previous chapters can be summarized using a variety of metrics (Table 10.1).

As evident from Table 10.1, the marine research partnerships explored in the preceding chapters use different partnership models of varying duration, different geneses and visions, and different funding regimes and implementation. Many of these partnerships were built from the bottom-up, i.e., researcher-driven (Nansen Legacy, COASST, Flooding Adaptation), others were top-down, i.e., funder-driven (MARES), some were a hybrid, i.e., the genesis of the Bering Sea, ARGO, and Deepwater partnerships were the result of simultaneous top-down and bottom-up processes coming together, while others have had elements of both during different phases of the program (Belmont Forum, MARINe). Depending on each of these situations the social processes involved entailed different activities and initial milestones. The issue of procuring funding in bottom-up partnerships could be addressed either *a-priori*, i.e., partners identify objectives or questions to be answered and attempt to gain support from funding sources, or *a-posteriori* in which case a more ad-hoc, often less integrated project, is started by already funded partners.

Despite different origins, approaches and visions, the narratives of the case studies have identified similar challenges, lessons learned, and key factors to which they ultimately attribute their success. Whereas in some sense one size does not fit all, a common adaptive process emerges that has allowed project champions to accommodate requests and information needs from multiple partners while keeping the collective focus on specific

Table 10.1 Summary information from all nine partnerships analyzed in this book.

Element/ partnership	Bering sea	Belmont Forum	Nansen legacy	Argo	MARES	COASST	MARINe	Deepwater (combined)	Flooding adaptation
Geographical location	Bering sea	Global	Barents sea and the adjacent Arctic ocean	Global ocean	Beaufort and Chukchi seas	Coastal Pacific Northwest (northern California, Oregon, Washington) and Alaska	US West Coast - Alaska to Baja California	Gulf of Mexico and NW Atlantic	Shishmaref, Alaska, Isle de Jean Charles, Louisiana
Duration (planning to closeout)	8y	7y - ongoing	10y	23y - ongoing	9y	23y-ongoing	31y - ongoing	8-10y	> 5y
# Of funding partners	5	16	12	~40	6	>50	5	3	1 (so far)
# Of research partners	32	79	10	>55	25	>20	20	8-16	9
# Of sectors	5	5	2	3	5	6	1	3	6
# Of people	>150	94	~240	>200	~80	6	250	~25-~50	~20
# Of nations	2	21	1	>55	2	1	4	1-4	3
# Of RFPs	3	1	1	Many (internationally)	1	>50	0	1	1
# Of awards per RFP (on average)	20	13	1	varied	1	~30	~30-50	1	1

Continued

Table 10.1 Summary information from all nine partnerships analyzed in this book.—cont'd

Element/ partnership	Bering sea	Belmont Forum	Nansen legacy	Argo	MARES	COASST	MARINe	Deepwater (combined)	Flooding adaptation
Cost (USD) to date— all sources	52 M	16.3 M	81 M	~28 M per year	10 M	$6 M	1.2 M per year	7–11.2 M	750,000
Policy implications	Fisheries management	Not yet fully captured	Outcomes drive new policies	Outcomes affect national policies particularly related to climate change	IARPC 5-year plan, NEPA	None	ESA, water quality, MPA, oil spill	Agency decision making for habitat protection	Potential to influence the visibility of non-academic researchers as co-PIs, climate adaptation policy implications
# Of disciplines	5	All	3	~6	5	3	2	>10	6
# Of domains	6 (land, sea, ice, atm, sediments, people)	6 (air, sea, land, ice, sediments, people)	4 (air, sea, land, ice)	3 (sea, ice, atmosphere)	4 (land, sea, ice, atm)	3 (biology = dead stuff on beaches; marine debris; people)	3 (land, sea, air temperature)	2 (ocean, sea floor)	1 (land)
Type of driver (research of researcher-driven funder-driven)	Hybrid	Both	Researcher-driven	Hybrid	Funder-driven	Researcher-driven	Both	Hybrid	Researcher-driven

Procurement types	Grants, awards, contracts	Grant, agreement	Grant	Grants, contracts, Cooperative agreements	Contract	Grants, contracts, MOUs, Cooperative agreements	Cooperative agreements, inter-Agency agreements, contracts, grants	Contract, verbal	Grant
Bridging Organizations involved	None	JPI-oceans Belmont Forum is a bridging organization	None	NOPP Internationally many	NOPP	None	None	NOPP	Shishmaref erosion and site expansion coalition, Kawerak
Goals	Ecosystem structure & function and impacts of climate change and fisheries	Pathways toward sustainable and equitable use of oceans, Accounting for and minimizing impacts of global change	Knowledge-based ecosystem management	Observe large-scale variability in global ocean temperature, salinity, ocean circulation and biogeochemistry	Ecosystem structure & function, ocean circulation	Nearshore ecosystem health	Monitoring rocky shores for the long-term application of biological science into policy.	Ecosystem structure and function, ocean circulation and vulnerability of sensitive biological components.	Understand relocation options for at-risk communities, understand multicultural responses to repetitive flooding

Continued

Table 10.1 Summary information from all nine partnerships analyzed in this book.—cont'd

Element/partnership	Bering sea	Belmont Forum	Nansen legacy	Argo	MARES	COASST	MARINe	Deepwater (combined)	Flooding adaptation
Vision	Climate change impact on marine environment and ecosystem uses, adaptation	Transnational, transdisciplinary approaches that leverage existing and planned resources	Holistic Arctic research project providing the integrated scientific knowledge base required for future sustainable management of the environment and marine resources	Global ocean observations for climate assessment, education, basic research, Blue economy, and hazard prediction applications	Integrated perspective of Arctic marine ecosystem that no partner could have achieved by itself.	COASST sees a future in which all coastal communities contribute directly to monitoring their local marine resources and ecosystem health through the establishment of a network of people and programs collecting rigorous and vital data.	To determine the health of the rocky intertidal habitat and make this information available to the public.	Bring together partners to address priority environmental issues that cannot be accomplished by a single agency or sector critical to understanding how sensitive biological systems function and elucidating the potential effects of human activities.	Understand the possibilities and limits to site expansion and relocation from a policy perspective and an Inupiat perspective. By comparing and contrasting these multi-epistemological perspectives, we expand the inventory of possibility to repetitive flooding.

Table 10.1 Summary of the nine partnerships presented, as case studies, in the previous nine chapters. Each of the nine partnerships represents a column in the table while each row represents a particular element of the partnership. These elements are defined as follows: **Geographical location** refers to the study area where observations were made, where the research took place; **Duration** includes the time it took from the initial idea through solicitation to project completion; **Number of funding partners** refers to the number of institutions or organizations that provided resources and support, including in-kind resources such as equipment or data; **Number of research partners** refers to the number of institutions or organizations involved in conducting research activities; **Number of sectors** refers to the type of organizations, e.g., private, public, academic, non-profit, military, tribal involved; **Number of people** refers to the number of people involved in the partnership, regardless of their role, while the **Number of nations** is the number of countries represented and this could include research and/or funding partners; **Number of RFPs** is the total number of requests for proposals, or solicitations, used to procure the research, and **Number of awards per RFP** is the total number of grants, contracts, or agreements resulting from those RFPs. **Cost** is the total amount of investments in US dollars, and this includes the value of in-kind resources provided. **Policy implications** refer to regional, national, or international policies that were drivers of the partnership and/or to policies that were written as a result of the findings generated by the specific partnership. The **Number of disciplines** refers to, e.g., physical, biological, chemical, and social sciences and their subdisciplines, while the **Number of domains** refers to air, land, ice, ocean, rivers, and sediments. **Type of driver** refers to the origin of the initiative and this could be funder-driven, researchers-driven, a hybrid, or both at different times. **Procurement type** indicates the type of binding instrument used to formalize the partnership, including contracts, grants, agreements, verbal, etc. **Bridging organizations** indicates whether a bridging organization was involved in facilitating the partnership of organizations, e.g., the National Oceanographic Partnership Program in the US. **Goals** and **Vision** refer to the goals and vision set in the early stages of the partnership.

objectives relevant to all involved. In this analysis, we first explore some of the key pillars of successful partnerships and then place these within the adaptive cycle of socio-ecological systems (Fath et al., 2015) to highlight key factors of partnership sustainability and resilience.

Pillars for success

Independent of the drivers that created the partnership most authors highlighted the importance of a shared vision, leadership, trust, flexibility, communication, and to some extent time, as key components. We characterize each of these in more detail.

Vision

Partnerships are about people, shared interests, and articulating a common vision that propels and keeps a partnership together. In some instances that may take the form of a conceptual model (e.g. Bering Sea, Nansen Legacy), a common desire to reveal unknown knowledge (e.g. ARGO, Deep-water Studies) and to bring together historically disparate players or geographies (e.g. Belmont Forum, MARINe), a need to inform regulatory processes with robust science (e.g. MARES), a vision to involve others in the scientific process toward a meaningful societal and environmental benefit (e.g. COASST, Flooding Adaptation), or any combinations thereof. The main point, however, is the shared desire of the partners to be part of something meaningful with a goal akin to their values. The "why?" articulated in this vision is the initial attractant as well as the sustained glue that holds the partnership together over time and in the face of challenges.

Leadership

Various leadership styles can be successful, but those that lead more toward consensus than top-down/directive seem to be most effective in forming and maintaining partnerships. The ability of leaders to keep inspiring all the partners toward the common vision is key for a maintained partnership and meeting the goals of the program. This idea of a "champion" (person, committee, or organization) requires trust and relationship building, and speaks to the benefit of consistency in leadership, even if not essential if adaptive processes are built into the partnership during the formative phase.

There is a nagging problem of how to distribute credit within a partnership, especially when promotions and future/continued funding for organizations or individuals are dependent on the ability to demonstrate clear successes and leadership. As Harry S Truman stated, "It is amazing what you can accomplish if you do not care who gets the credit."

In the nine partnerships examined in this book it was clear that whereas at least one champion (organization, committee, or person) with the will and motivation to clear all obstacles was one of the common denominators, the partnership was successful because of the collective accomplishments.

Trust

Without good leadership is difficult to develop trust and establish a sustainable partnership. The partners (people) need to establish trust or have some history that buttresses their relationship and take actions that stand behind their words. Building trust takes time, which is why in most cases explored in this book, partnerships were built on a core of existing relationships and trust, allowing for others to be integrated without destabilizing the core dynamic. Sustained relationships and trust thus involve the use of other resources and resilience properties such as time, leadership, team dynamics, and communication, which facilitate flexibility and connectivity from all involved.

Trust-building may also include processes of formal recognition of historical injustices or being frank about the relative financial, cultural, and social positioning of the partners. When working with communities impacted by climate change or marine management, for example, it can be important to recognize that outcomes of research and research partnerships frequently have immediate and personal implications – making these projects and the knowledge created within "more than data" to some participants (Stoecker, 2009; Adams et al., 2014). Risk-taking for senior researchers or scientists that have permanent positions within academic institutions, government agencies, or industries, may be easier than for scientists on soft money, early career researchers, or scientists who are on the job market (Hein et al., 2018). Indigenous partners may carry additional responsibilities of being cultural translators and ambassadors for research projects – which is a risky and weighty burden, given histories of research as an extractive and colonial process (Cochran et al., 2008). Trust, as we have learned in these chapters, takes time and is central to sustainable partnerships. Trust building may include, and increasingly will be expected to include, being forthright about the benefits and risks each partner takes when entering into a research endeavor.

Flexibility

Effective flexibility, or the ability to adapt to changes during the creation and implementation of the partnership and project while staying true to the overall vision, is in many ways a resultant characteristic of effective leadership, established trust, and clear and transparent communication. A ten-year plan is a wonderful thing; following it is quite another. Technology changes, partners change, needs change, budgets change, so being flexible and adaptive is essential. One good purpose of a ten-year plan, however, is, under the assumption of no uncertainty about the future, to show how it all fits together (vision), how actions meet needs, and budgets represent reality. After all, if the plan doesn't make sense and meets constraints under the best of assumptions, how can it possibly expect to be worthwhile when changes start taking place? A good ten-year plan (or five or even three) not only explores the future, but offers contingency plans (for budgets, personnel, logistics, etc.) should each of the major assumptions be in error. The goal is for future change to not be an emergency, but just an annoyance, if not an opportunity, where all partners work together to compromise, solve, and adapt.

Communication

Successful partnerships depend on effective (clear, frequent, useful) internal and external communication. Without it the vision is not clear, leadership is ineffective, trust can be eroded, and flexibility is hampered. Successful partnerships have established internal communication plans or processes to facilitate frequent interactions among and between all key players (funders, project personnel) to keep the project on track (objectives, schedule, and budgets), facilitate constant integration, avoid misunderstandings, and communicate changes where needed. Plans for external communications are equally important to be able to highlight the benefits of the project to funding or implementing organizations, as well as to convey findings to decision-makers and society at large. This forces clarity in deliverables (beyond technical peer-reviewed papers and conference presentations) and helps ensure that results are used in the real world, while also garnering support for sustained support of the current partnership or future endeavors.

Time

Although not explicitly stated in most case studies, the issue and importance of "time" needs to be highlighted. Relationship and trust-building, the most important components for partnership building and sustainability, take time. The development of shared goals takes time. Getting institutional buy-in (the partners are people, but it is their institutions that support them and back them) takes time. Even after the formation of a partnership, getting results takes time, getting exposure and application of those results takes time, as does the realization of the value of these partnerships by funding agencies and traditionally uni-disciplinary scientists. Years to develop a functioning partnership is not unusual. From Table 10.1 we can appreciate that in all cases time investments involved significant dedication and resources. To be successful, it is important to take the time to do this right, rather than rushing a process that is not ready. As Albert Einstein said:

> It's not that I'm so smart, it's just that I stay with problems longer.

In brief, partnerships can originate driven by the need to use outcomes, from the desire to learn about the natural world, or to challenge the development or advancement of new technologies. Regardless of how and why they emerge, we have identified a suite of fundamental properties and resources that will enable sustainable partnerships and ultimately help reach the goals originally set.

Resilience and sustainability of partnerships

Aside from the key pillars outlined above, the nine case studies of partnerships presented in this book have shared more specific details on the elements and processes that have contributed to their successes, including their sustainability over time. To put this into a broader context and allow for a framed analysis of key partnership elements, we define Partnership Sustainability as the product of *Resilience* and *Resources*, where Resilience is a necessary but not sufficient condition for Partnership Sustainability, *and* where it is assumed that the necessary Resources are accessible.

Partnership Resilience in this case is characterized by four contributing factors (adapted from Auad et al., 2018): *Connectivity, Flexibility, Redundancy, and Diversity* (Fig. 10.1).

These contributions were present in most of the partnerships presented, and were identified by their authors as e.g., communication and trust (connectivity), adjusting to unforeseen events (flexibility), verification and validation processes for quality control (redundancy), and including scientists and knowledge holders from multiple disciplines (diversity).

Fig. 10.1 Elements of sustainable research partnerships.

The nine case studies also allow us to provide a list of ten important Resources that support resilient and sustainable partnerships. These resources span social, economic, physical, and electronic elements. Some of these were already discussed as *Pillars for Success* above, but all contribute to sustaining partnerships over their life cycles, in some cases, over decades. These resources are *Time, Funding, Leadership, Knowledge, Team Dynamics, Information* (including data), *Technology, Creativity, Intuition,* and *Vision* (Fig. 10.1).

It is important to note here the complex interplay of some of these resources. For instance, quality leadership and a healthy team dynamic would certainly favor building relationships and therefore trust, which are all needed elements for successful team performance (Vilar et al., 2012). The latter could in turn be considered one important factor to positively hardwire the social system under consideration in a resilient fashion.

Following this line of thought, we consider partnerships as a social complex system where multiple interactions of its multiple elements can exist at any point in time within an adaptive cycle, noting that resilient systems (in this case research partnerships) are those that successfully navigate all stages of this cycle: new beginning, growth and development, conservation or status quo, crisis and collapse, and reorientation and reorganization (Fath et al., 2015). This translation of research partnerships to the complex systems arena is useful, because analysis tools, as well as numerous methodologies, have already been developed for such a framework. In practice, this approach can be used for building resilience in socio-ecological systems that involve multiple resource managing entities with overlapping and/or complementary geographical jurisdictions, as is common in adaptive governance frameworks (Auad and Fath, 2021). In addition, the creation and implementation of partnerships have been proposed as a potentially effective way to mitigate negative socio-ecological impacts (Wiese, 2021).

The identification of resilience components and resources does not mean that all are needed or are equally critical to supporting the partnership all the time. For instance, the availability of previously acquired information and knowledge (scientific and/or traditional) can shed light on social issues, biophysical processes, or anything else that helps the partners to plan the development of new understanding. Conversely, technology or new information may be relevant, but not critical, during a crisis, whereas leadership would be.

By examining the nine case studies in the previous chapters we can identify when a particular resource or a component of the overall resilience of the partnership is needed the most (critical); similarly to the analysis presented by Fath et al. (2015) for business groups who have common and/or complementary goals. In Fig. 10.2, we visualize the components of resilience and resources to the four stages of the partnership adaptive cycle (building on the initial phases identified by Holling (1986)).

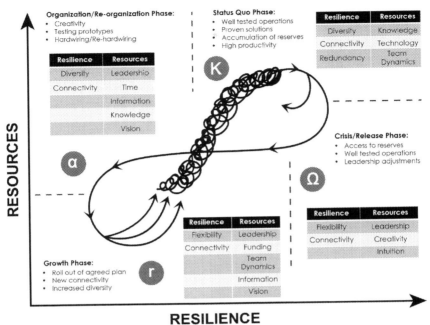

Fig. 10.2 Resilience characteristics and resources deemed critical within the different stages of the partnership adaptive cycle.

Conclusions

The specific success and barrier factors for each of the four phases are mentioned in nearly all of the case studies, and reflect an amalgamation of material previously reported in the literature (e.g., McKendall, 1996; Willey, 1999; Mattessich et al., 2001; Carnwell and Carson, 2005; Briscoe, 2008). Either by design or circumstance, all nine case studies presented had the necessary resilience characteristics and resources available at the right times to make their programs successful and sustainable. Some of their goals were related to ocean sustainability, others explored additional horizons such as citizen science, co-production, and resilience of socio-ecological systems. Regardless, future partnership efforts should carefully examine these key characteristics and components to enable their successes.

It is worth noting here that if one of the partners has operational goals (e.g., regulatory), bridging organizations contribute to a more adaptive approach to decision-making. These bridging organizations (defined as formal organizations that use specific collaborative mechanisms to bring together diverse actors) have been shown to have a crucial role in implementing adaptive governance strategies (Brunner et al., 2005) as a key element to managing for (Djalante et al., 2011) and increasing resilience by augmenting the quality and quantity of connections in the system. Several partnerships examined in this book (MARES, Argo, Deepwater Partnerships, Flooding Adaptation) acknowledged the fundamental role played by bridging organizations such as the National Oceanographic Partnership Program (NOPP), while the Belmont Forum itself could be regarded as a bridging organization or a partnership of partnerships.

Finally, we cannot overlook the importance of ethics in driving large integrated multi-, inter- and trans-disciplinary studies that can generate a higher level of understanding, often useful in decision-making across sectors and nations. It is this consideration for ethical outcomes, chiefly environmental sustainability, that has driven scientists, managers, and policy-makers to provide increasing support for larger and more comprehensive scientific research activities and projects. The social process of partnering supports those ethical visions and goals even if a collection of factors keeps us far from being efficient. These include the traditional mono-disciplinary approaches used in educational institutions, as few universities include integrated, interdisciplinary courses and subjects in their curricula; the silo-like structure of many government agencies; as well as other well-known pathologies of decision-making such as fragmentation and groupthink (Ramirez and Wilkinson, 2016). Some of these may be resolved by creating greater awareness about collaborative benefits like we have attempted to do here; others may be addressed by reexamining institutional research policies. We address the latter in the next chapter.

Acknowledgments

The views and opinions expressed in this chapter are those of the authors and do not necessarily reflect the official policy or position of their employers or any other agency or organization.

References

Adams, M.S., Carpenter, J., Housty, J.A., Neasloss, D., Paquet, P.C., Service, C., Walkus, J., Darimont, C.T., 2014. Toward increased engagement between academic and indigenous community partners in ecological research. Ecol. Soc. 19 (3).

Auad, G., Fath, B.D., 2021. Recipes for a flourishing Arctic. In: Wassmann, P. (Ed.), Whither the Arctic Ocean? Research, Knowledge Needs and Development en Route to the New Arctic. Fundación BBVA, Bilbao, Spain, pp. 75−86.

Auad, G., Blythe, J., Coffman, K., Fath, B.D., 2018. A dynamic management framework for socio-ecological system stewardship: a case study for the United States Bureau of Ocean Energy Management. J. Environ. Manag. 225, 32−45.

Briscoe, M.G., 2008. Collaboration in the ocean sciences: best practices and common pitfalls. Oceanography 21 (3), 58−65.

Brunner, R.D., Steelman, T.A., Coe-Juell, L., Cromley, C.M., Tucker, D.W., Edwards, C.M., 2005. Adaptive Governance: Integrating Science, Policy, and Decision Making. Columbia University Press.

Carnwell, R., Carson, A., 2005. Understanding partnerships and collaboration. In: Effective Practice in Health and Social Care, pp. 4−20.

Cochran, P.A., Marshall, C.A., Garcia-Downing, C., Kendall, E., Cook, D., McCubbin, L., Gover, R.M.S., 2008. Indigenous ways of knowing: implications for participatory research and community. Am. J. Publ. Health 98 (1), 22−27.

Djalante, R., Holley, C., Thomalla, F., 2011. Adaptive governance and managing resilience to natural hazards. Int. J. Disas. Risk Sci. 2 (4), 1−14.

Fath, B.D., Dean, C.A., Katzmair, H., 2015. Navigating the adaptive cycle: an approach to managing the resilience of social systems. Ecol. Soc. 20 (2).

Hein, C.J., Ten Hoeve, J.E., Gopalakrishnan, S., Livneh, B., Adams, H.D., Marino, E.K., Susan Weiler, C., 2018. Overcoming early career barriers to interdisciplinary climate change research. Wiley Interdiscip. Rev. Clim. Chang. 9 (5), e530.

Holling, C.S., 1986. The resilience of terrestrial ecosystems: local surprise and global change. In: Clarkand, W.C., Munn, R.E. (Eds.), Sustainable Development of the Biosphere: Interactions between the World Economy and the Global Environment, vol. 14. Cambridge University Press, Cambridge, UK, pp. 292−317.

Mattessich, P., Murray-Close, M., Monsey, B., 2001. Wilder Collaboration Factors Inventory. Wilder Research, St. Paul, MN.

McKendall, V.J., 1996. Factors Facilitating Interorganizational Collaboration. PhD Dissertation. University of Minnesota.

Merton, R.K., 1973. The Sociology of Science: Theoretical and Empirical Investigations. University of Chicago press.

Ramirez, R., Wilkinson, A., 2016. Strategic Reframing: The Oxford Scenario Planning Approach. Oxford University Press.

Stoecker, R., 2009. Are we talking the walk of community-based research? Action Res. 7 (4), 385−404.

Vilar, L., Araújo, D., Davids, K., Button, C., 2012. The role of ecological dynamics in analyzing performance in team sports. Sports Med. 42 (1), 1−10.

Wiese, F.K., 2021. Why did the Arctic not collapse? In: Wassmann, P. (Ed.), Whither the Arctic Ocean? Research, Knowledge Needs and Development en Route to the New Arctic. Fundación BBVA, Bilbao, Spain, pp. 229−238.

Willey, T., 1999. Overcoming Barriers to Collaborative Research: Report of a Workshop. National Academy Press, Washington, DC, p. 60.

CHAPTER 11

Research partnerships and policies: a dynamic and evolving nexus

James J. Kendall, Jr.[a], Elizabeth K. Marino[d], Melbourne G. Briscoe[c], Rodney E. Cluck[f], Craig N. McLean[e] and Francis K. Wiese[b]

[a]U.S. Department of the Interior, Bureau of Ocean Energy Management, Anchorage, AK, United States; [b]Stantec Consulting Services, Inc., Anchorage, AK, United States; [c]OceanGeeks LLC, Alexandria, VA, United States; [d]Oregon State University - Cascades, Bend, OR, United States; [e]National Oceanographic and Atmospheric Administration, U.S. Department of Commerce, Silver Spring, MD, United States; [f]U.S. Department of the Interior, Bureau of Ocean Energy Management, Sterling, VA, United States

Introduction

In this chapter we examine how research policies influence the creation, sustainability, or deterioration of research partnerships. To focus this discussion, we define research policies in terms of input and output: the input is any legislation, regulation, administrative action, incentive structure, or normative condition; the output is the influences of the policy on how research is created, funded, administered, carried out, and even remembered.

In the case studies presented in this book, research policies are an influencing condition that partnerships depended on, had to work around, or a combination of both, to achieve their goals. We first discuss the general nature of policies and partnerships, and how such policies can influence the intended research. We then briefly look at how policies interacted with the case studies, and finally we focus on how existing research policy frameworks may evolve to efficiently and equitably foster future partnerships capable of tackling the challenges of the 21st century. While recognizing that many of these examples are from a US perspective, we hope that these generalized research policy considerations and associated social processes will also serve a more general purpose and support national and international marine science partnership efforts such as those described in the Argo, Nansen Legacy, and Belmont Forum case studies.

Research policies and partnerships

One way to view policy is as an authoritative statement of principles that guide decisions and actions toward the attainment of the desired outcome. A policy document may contain implementing rules to guide the conduct of those individuals and activities covered by the policy-defining authority, which could include a government, corporation, community, or other organization. Thus, a policy is both a notion of principle or value and can also be a written document implementing the same by defining practices that guide conduct. In the implementation stage, policies can also be subject to interpretation – the latitude within which written policies may be interpreted – which we understand as authoritative discretion. Both written policies and their latitude of discretion may structure what research partnerships are possible.

Once formed, research partnerships become social contracts whereby a group of people or institutions come together to achieve a common outcome that could not be achieved without collaboration as discussed by Briscoe (2008) and as articulated in the enabling legislation of the National Oceanographic Partnership Program (NOPP, 1997). NOPP was established by the US Congress (Public Law104-201) in 1997 to coordinate and strengthen our oceanographic efforts by identifying and carrying out partnerships among federal agencies, academia, industry, and other members of the oceanographic scientific community in the areas of data, resources, education, and communication. On January 1, 2021, Congress reauthorized and strengthened NOPP as an amendment included in Section 1055 of the National Defense Authorization Act for Fiscal Year 2021. See Public Law 116-283.

Historically, academic research partnerships were focused on advancing scientific knowledge through relationships between scientists, sometimes across disciplines, while leaving other experts (e.g., Indigenous knowledge holders, funders, and end-users) on the sidelines. One outcome of this was the relative lack of focus regarding the impacts of scientific inquiry on society. However, as discussed in the Introduction and demonstrated in the case studies, the concept of partnership has evolved taking on a broader, more inclusive, and arguably richer perspective. Research partnerships today are frequently inclusive of stakeholders who may be impacted by, or benefit from, scientific inquiry, such as end-users, communities, businesses, non-governmental organizations, and governments. They can also include collaboration with local knowledge holders, such as partnerships between scientists and commercial fishing experts or farmers that have experiential knowledge about animals, weather, or climate. Research partnerships can also be a collaboration among different epistemological experts, who are brought up within, and have the expertise that stems from, different knowledge systems. An example of this is when scientists partner with Indigenous knowledge holders or other experts within diverse knowledge systems.

Previously, scientific questions were primarily addressed by examination through the scientific method, whereby scientists define a question, design a process of investigation (e.g., an experiment), collect data, analyze results, and communicate their findings through peer-reviewed publications, at conferences, or among other accepted means. Although it can be iterative on occasion, it is typically a methodologically linear progression for most scientific investigations and shares components with other systems of knowledge production and thought (Aikenhead and Ogawa, 2007). Today, however, research questions come not just from scientists, but from a range of knowledge holders and citizens who often have the most to gain or lose in the application of research findings. Likewise, the concept of "expert" has expanded to include expertise that is developed among traditionally non-academic knowledge systems, or people whose experiences or occupation give them insight into the way a social, ecological, or economic system functions.

Bringing together and using different knowledge systems into the knowledge production process is showing great promise (ABR et al., 2007; Coleman et al., 2012; Galginaitis 2014a,b; Kendall et al., 2017; Gryba et al., 2021). If scientists are interested in knowledge building and intellectual innovation, then the moments in which one set of epistemological limits are reached, and then breached by the epistemological limits of another, are critically important. For example, the limitations of government policy to respond to climate change outcomes are particularly illuminated when compared to Inupiat responses to climate change. Putting these knowledge systems side-by-side is useful in understanding how adaptation possibilities are limited by our prevailing practices and imaginations, but it is also challenging. For example, it has typically been viewed that stakeholders should not be peer reviewers and peer reviewers should not be stakeholders. Under the broader view of partnerships presented in this chapter, such a policy may be increasingly challenged.

Within partnerships, there are foundational and complex questions of how to give equitable respect and attend to all participant protocols and norms (Whyte et al., 2016). While addressing power disparities and power dynamics is particularly important when different knowledge systems are brought together, acknowledging the potential for power disparities is important across most research partnerships. Who is in charge of resources? What accolades and accomplishments are recognized within institutional structures? Who qualifies for overhead costs? How are new, non-traditional partners compensated? What is considered legitimate data? What is considered a legitimate product? Understanding these dynamics are structural and policy issues that should be addressed to achieve *equity* across partnerships. This realization may also be where updated research policies could greatly enhance partnership possibilities and modes of behavior.

These concepts are evolving, and these questions are being asked at a time when scientific and governmental institutions are sometimes being challenged by the public. Examples of this could include the levels of trust in these institutions on such topics as climate change and how best to address a pandemic. How do we open science to a wider range of participants and decolonize it to the extent that it is no longer seen as the only legitimized way of coming to know, without losing the capacity of scientists to function as critical knowledge producers who create and hold information needed to address the societal and ecological challenges of the 21st century? By "decolonize" we refer to the well documented inquiry into the set of "scientific and pseudo-scientific" practices and institutions that helped to promote the colonization, or the European occupation, of Asia, Africa, South America, and North America. These racist practices included identifying who was a legitimate "knowledge holder" or "scientist" and continued to persist well into the 19th and 20th century. Social scientists in the 1960s and 1970s began to investigate science and education as a subject of research more broadly and noted the power dynamics and historical contingencies therein (Freire, 1968; Latour and Woolgar, 1979). Contemporary critiques of research in Indigenous contexts, and the possibility of reclaiming alternative methods of "coming to know" (while having always been practiced), emerged more visibly with the 1999 publication of *Decolonizing Methodologies: Research and Indigenous Peoples* by Linda Tuhiwai Smith (1999). We use the term "decolonize" here to acknowledge the colonial positionality of science as an institution for the last 400+ years, and to facilitate the understanding, acknowledgment, and dismantling of racism, patriarchy, and colonialism within the scientific process, which is necessary to form equitable partnerships. Furthermore, we also put forth the question: how can our understanding of research partnerships help expand the types of knowledge we include in decision-making, expand the role of non-scientists in a scientific project, and yet solidify and reiterate the need for science and scientific inquiry?

Case studies

The case studies in this book document a few salient takeaways regarding research policies. First, research policies that allow for significant time and effort to create shared objectives are required to begin and sustain partnerships. In the MARES project (Chapter 5); the Bering Sea Project (Chapter 1), and the climate change relocation project (Chapter 9), long-standing relationships were necessary for success: these were initiated informally before project implementation, as part of the pre-planning for a research proposal, or as part of the research project itself. Policies that fund scientists and other knowledge holders, within or outside institutional structures, allow for the time needed to create robust partnerships are important.

Second, potential partners are more likely to engage if: (1) the approach is familiar; (2) the process is open and transparent; and (3) the process has limited requirements for participation (e.g. matching resources). In some cases, formal arrangements were key to partnership sustainability. Formal arrangements and internal policy mapping were particularly important when the partnership included multiple government agencies as in the Belmont Forum (Chapter 2). However, in instances when small private businesses or previously established citizen-science groups were science partners, as in Chapter 6, informal and non-arduous requirements for participation allowed for the sustainability of loosely affiliated, but engaged, community groups.

An example of a partnership originating informally is documented in Chapter 7, where no written documentation in the form of a Memorandum of Understanding or Interagency Agreement existed between the funding entities. The relationship was entirely founded on the trust established by a long-term working relationship then followed by the historical associations between many of the principal architects of the partnership. These attributes resulted in a collective focus on the goal, not on the impediments. This rapidly led to a spirit of collaboration among the government funding entities, the private sector, managers, decision-makers, and an international, interdisciplinary team of scientists focused on success. Trust, familiarity with the other potential partners, drive, and flexibility all played a role in developing a successful world-class partnership. Likewise, while scientist participants are often required to meet their scientific goals, the guiding research principles or policies acknowledging that the benefits of any partnered endeavor must bring something meaningful and tangible to all partners involved, were critical to long-term sustainability — particularly among non-science partners. Internal policies that focused on flexibility in what constitutes the necessary components of a project, or on what constitutes product development, were important to multiple projects.

Third, policies that allow for easy international, cross-agency, and cross-sector collaboration were critical, and in some cases, were created to promote partnerships. Some organizations and research policies specifically focused on partnerships that already exist; one such international body is the Belmont Forum, established in 2009 as a partnership among funding organizations, international science councils, and regional consortia. Another example is the US National Oceanographic Partnership Program established in the 1997 National Defense Authorization Act (Public Law 104-201), and reauthorized in the 2021 National Defense Authorization Act (Public Law 116-283) to facilitate partnerships among US federal agencies, academia, and industry to advance ocean-science research and education. On a more regional scale, the US Energy Policy Act of 2005 established the North Slope Science Initiative (NSSI) to better inform management decisions about industrial development on the North Slope of Alaska. With a broad legislative mandate, NSSI was intended to integrate research across federal, state, and local governments and to increase partnerships with industry, academia, non-governmental organizations, and the public. Multiple case studies noted that these large-scale partnerships were relatively novel 20 or 30 years ago but have grown more popular over time, in part because of their success and the comradery which developed among this generation of scientists. Regarding the latter, one direction to consider is how to create large-scale, multi-sector partnerships that include early-career and less established scientists as well as smaller businesses, community groups, or not-for-profit institutions.

Fourth, policies that allow for flexibility when moving funding, personnel, and data within partnership organizations, and policies that allow for flexibility with timeframes and deadlines, will increasingly be useful. This was true of projects in this book regardless of scale. When diverse sets of agencies, actors, sectors, and cultures come together, there is certain to be a substantial amount of uncertainty. This was brought to the forefront during the COVID-19 pandemic when all aspects of partnership-building and methodological plans were disrupted. Among partners targeting sustainable projects, flexibility, improvisation, and creativity are all needed in the face of uncertainty or disruption; but even more important are iterative learning and preparedness in the overall pursuit of sustainability of the partnership.

Policy sources and impacts

Policies affecting partnerships can be generated from outside the partnership (i.e., funders, legal requirements, university requirements, statutory missions, etc.) as *exogenous* sources, or they can be generated by the partnership itself, internally, as an *endogenous* source. The policies themselves might be helpful to the partnership or be limiting (in whole or in part) to the partnership building, the number of partners, their commitments, and therefore the scope of a given project. This spectrum of research policy types and their impacts on partnerships is shown in Fig. 11.1. When considering this model, however, it must be noted that flexibility is key due to shifts in perspectives and on a case-by-case basis. For example, "procurement type" is a key tool that can be limiting or helpful. Significant effort can be invested from the funding perspective in determining how to best make an opportunity available to potential partners. It can be exogenous and endogenous in that the funding organization procurement vehicle

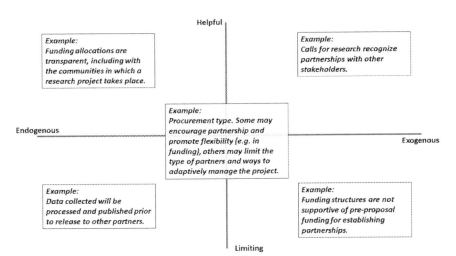

Fig. 11.1 Policy sources and impacts, with examples.

is clearly exogenous, but the type of agreements between the partners that move resources from one to another (e.g. as subcontractors) is endogenous. Characteristics of both are extremely important and impact the flexibility and reporting requirements by all involved.

Once a project is underway, the partnership can do little to change an exogenous and limiting policy (lower right in the figure), and simply must learn how to work with it or around it. Funders and agencies, and those who make and promulgate such policies, should be wary of any of their policies that fall in this unfortunate quadrant and consider deleting or revising them to be less limiting and more helpful. Individuals within the partnership have control over its endogenous policies, so any that are limiting a viable and productive partnership should be carefully examined as to their purpose and necessity.

Another example might be the US Federal Acquisition Regulations (FAR) which often appear to limit the development of partnerships with the private sector; this is an extreme exogenous and limiting policy for the formation of partnerships. While public/private sector partnerships are encouraged and policies are created to promote this engagement (e.g. NOPP), the limiting factors can be paramount. In some cases, the burden of developing the partnership could overshadow the benefits.

The middle ground in Fig. 11.1 is meant to be external to the partnership, but less exogenous than coming from, say, the funding entity. The example given is the procurement type. Different procurement vehicles (e.g., contracts, cooperative agreements, grants, interagency agreements) have different pros and cons regarding funding, integration, management, reporting requirements, new partners, and more. In other words, different procurement vehicles imply different flexibilities on different fronts, both from the initial funding entity as well as between the lead organization conducting the work and its partners. Budding partnerships should carefully consider the implications of these not-quite-exogenous policies, with an eye toward promoting characteristics that support flexibility in partnerships as described in Chapter 10 (lessons learned) and expanded on below.

The nine case studies in this book have all been subject to policies that reside in all the quadrants of the figure. Our goal is to encourage partnerships being formed to look carefully at the policies under which they will function. This would include, for example, spending time pre-planning and expending some effort to examine what might work for them and what might work against them.

Finally, there are research policies that may advantage, enable, and be helpful to some partners while at the same time be limiting to others within the same project. This aspect of research-policy making complicates an easy assessment of whether a policy is limiting or helpful. The ongoing assessment of the demographic make-up, social use, and cultural embeddedness of research will help us understand and encourage policies that engender greater diversity in scientific and research-based projects.

The future of research policies

At the highest level of analysis, the relationship between research policy, knowledge production, end-users, and societal impact is a foundational social contract and can be viewed itself as an ultimate, if frequently unacknowledged, global partnership involving:

- the personal or human relationship among the individuals gathering and/or producing new knowledge;
- the recognition that producing knowledge, by its very nature, is a sociocultural process;
- the societal impacts of producing knowledge that will then often influence future knowledge production (i.e., new questions are discovered); and,
- the fact that new knowledge may lead to the development of new tools (e.g., policies themselves, and to an influence on decision-making at multiple scales).

Ideally, research policies ought to be crafted to support, or at least not limit, these processes. If partnerships are viewed as existing upon the backdrop of society rather than apart from it, then any product of a knowledge provider is a contribution to society, with some products of direct, immediate applicability, often built directly into the relationship from the onset (Fig. 11.2). All these products rely on some form of relationship, be it a simple use of societal support such as a grant to a single expert, a network of experts, or a fully collaborative effort involving a team of participants as defined in the Introduction of this book and as documented in the case studies.

As shown in Fig. 11.2, all partnerships exist upon the backdrop of society where any product of a *Knowledge Provider* is a contribution to society, with some products possibly of direct, immediate applicability; the latter can be built directly into the relationship (e.g., applied science). *Funders* are

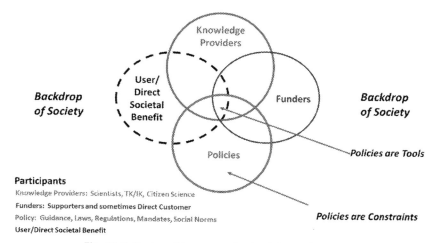

Fig. 11.2 Partnerships upon the backdrop of society.

those members of the relationship supporting knowledge providers; in relationships involving applied science, a funder may also qualify as a *User*. *Policies* are instruments that either facilitate or limit the desired relationship (see Fig. 11.1). Where the policies facilitate the relationship and are helpful, they can serve as *tools*; where they harm and limit the relationship, they are *constraints*. The optimal, and often the most sustainable, relationship is where policies facilitate, and products benefit, each participant and society in general.

As we look toward encouraging future research policies that address participants, funding, recognition, compensation, and communication of outcomes, it is helpful to remain cognizant of both the *endogenous* and *exogenous* components (Fig. 11.1), and the continuing evolution of the social landscape upon which partnerships are formed. Regarding *endogenous* components, partnerships depend on the efforts of the participants which includes an acknowledgment, first and foremost, that this is hard work and this work is a legitimate part of the endeavor. This will always be the case, even with the existence of policies that aggressively facilitate, promote, or even reward the formation of partnerships. It all starts with the participants (people) and their relationships. Regarding *exogenous* components, also to be considered are:

- situational awareness of such matters as laws, regulations, executive orders, etc.;
- attentiveness, including open-mindedness, to such matters as non-traditional partners;

- power dynamics, social media, transparency;
- knowledge users; and
- local, national, and international politics.

The last bullet contemplates being politically astute. Here, being politically astute refers to a general willingness to: understand the interrelationships and responsibilities of working with multiple and diverse individuals and organizations; understand and acknowledge the priorities, cultural norms, and potentially unwritten rules of potential partners; consider the perspectives of potential partners in making decisions and the potential ramifications thereof; demonstrate sensitivity when forming partnerships with non-traditional partners; and, understand the geo-political environment relevant to the topic of the intended work (e.g., changing climate, a pandemic, etc.).

Following this general awareness, an ideal set of helpful *exogenous* and *endogenous* research policy characteristics upon which to build strong and sustainable partnerships should be based on the following foundational principles or considerations.

(1) **Clarity of the intent of partnership:** What is the desired, specific knowledge outcome? For research, for the most part, delivery of the end products (e.g., data, results, published reports, and papers) has been the norm. Policies should clearly articulate their intent and purpose concerning partnerships.

(2) **Institutional Goals:** The goals of participating institutions should be clearly defined. It must also be stated whether the entity is a funder, research participant, end-user, or mixture.

(3) **Flexibility and Transparency:** Address how resources are distributed with an emphasis on flexibility.

(4) **Flexibility for Expansion:** Often, once an effort has been underway, the need for a legitimate expansion in terms of personnel, funding, logistics, resources, etc. comes to light. Policies should be crafted to acknowledge such possibilities and not be purely focused on the prevention of mission "creep." This dovetails with "trust" as well as the sustainability of successful partnerships discussed in Chapter 10.

(5) **Recognition of the legitimacy of different systems of knowledge:** Policies that allow for, and even facilitate, the inclusion of researchers, knowledge holders, etc. that do not fit the classic research model. This would include Indigenous knowledge, citizen science, crowdsourcing, and the like. These policies must also address how such participants can be compensated. For example, scientists going

to the Arctic are paid; however, village elders in Arctic communities sharing their Indigenous knowledge are often not paid.

(6) Recognition and acceptance of different products: For many former and present-day research partnerships, peer-reviewed documents are considered a desirable product. However, with the inclusion of different systems of knowledge (#5), citizen science, and a much more diverse participant list, new metrics of success will have to be used.

Although addressing research policy needs may appear daunting, part of this change is already in process. Over the past few decades, integrated, interdisciplinary scientific investigation has been increasing and, recently, the combined use of science and Indigenous knowledge to address resource management questions is becoming more and more common — almost to the point of being expected. As such, scientists and knowledge holders can all take an active role in the research policy formulation.

The profile of research has changed from the concentrated classical silos to the interdisciplinary integration of even non-traditional subjects, constituents, and participants. The policies controlling or encouraging research should seek to recognize this transformation and account for the opportunities of innovation and discovery, not the control and confinement of intellect and ambition to comport with an administrative historic norm. The partnership participants should plan thoroughly and consider possible solutions to future events before they arise, perhaps by using the methods of "scenario planning." Today's scientist is far more interdisciplinary in training; information is more abundant and available; and the results, in many cases, are breathtakingly rapid. Sponsors and funders are increasingly seeking the societal relevance of the research. Research policy and performance must comport with these changes, from the endogenous outlook of the partnership participants to the exogenous guidance of the sponsoring or funding parties. Expanding the tether and scope of partnership policies, and addressing the recommended research policy characteristics above, will help ensure the longevity of the partnership, unity, diversity, and equity of the partners, success of the project, and relevance and communication of its results.

At the time this chapter was prepared, two topics garnered worldwide attention: the changing climate and the development of pandemic vaccines. Research in these areas required a worldwide effort and research policies ought to foster the partnerships needed to tackle these acute and chronic challenges. Indeed, the importance of strengthening multi-stakeholder partnerships and the recognition that we need to encourage and promote effective public, public-private, and civil society partnerships, building on the experience and resourcing strategies of partnerships, is at the core of United Nations Sustainable Development Goal 17. This is generating significant momentum at a global scale including requests for proposal that require consideration of some of the 17 sustainable development goals. The Belmont Forum, for instance, has been addressing this through several of their recent requests for proposal.

Creating new or changing existing research policies to make them more helpful will require an "all hands-on deck" approach and consideration of the societal landscape. It will also require a high level of political astuteness and the involvement of a wider array of participants. If we are to tackle today's and tomorrow's socio-ecological grand challenges, we will all need to work together. As knowledge progresses and evolves, so must the processes and tools to support it.

Recommendations for research policies

To make progress on building strong sustainable partnerships for the 21st century, we propose the following recommendations pertaining to research policies as discussed in this Chapter. We believe these will foster and promote additional, and more robust, sustained partnerships.

1. Continuously review existing research policies and adapt and formulate new policies that enable and support partnerships; particularly among partners who have previously been excluded from scientific endeavors.
    a. To find novel ways to support formal and informal partnerships and prevent hindering these relationships.
    b. With input from all partnership participants, work to identify structural challenges within regulatory and policy frameworks.
        i. Identify when these obstacles are the results of law, internal discretion, or procedural history; and, concurrently,
        ii. identify where flexibility may exist and encourage research funders and other gatekeepers to enact discretion in flexible and inclusive ways.
    c. Identify and encourage highly motivated individuals to engage policy makers in implementing needed policy changes. Such individuals should be acknowledged for their efforts by all participants (i.e., knowledge providers, funders, policy experts, and society in general).

2. Develop policies that instill the value of partnerships early in professional training and early-career development programs. Early-career scientists should be encouraged and evaluated on:
   a. The quality of their work and how and to what extent they engaged with others; and,
   b. How well their work is situated within history and society. As we described earlier, partnerships are social contracts between scientists and others who do the work, and society which receives the benefits of that work.
3. Seek award and recognition policies that foster a wider institutional appreciation of the value of partnerships and the work results beyond (not instead of) the traditional peer-reviewed literature and professional conferences.
   a. Better document societal benefits and outreach results.
   b. Social media can be a valuable tool in disseminating the benefits of the work conducted and the partnerships employed, while also increasing transparency and legitimizing the resources invested.

We suggest these recommendations will do much in encouraging and sustaining more robust and diverse partnerships. The goal of good research policies is to move us toward better collaboration—the highest level of partnership—leveraging our resources, and improving diversity, equity, and inclusion in the production of knowledge for the 21st century.

Acknowledgments

The authors of this chapter wish to acknowledge the constant encouragement and guidance provided by Dr. Guillermo Auad. While his contribution warranted that of co-authorship, he insisted on remaining in the background even though he was instrumental in bringing this chapter to successful completion. The authors would also like to acknowledge Ms. Shannon Vivian for providing an editorial review. The views and opinions expressed in this chapter are those of the authors and do not necessarily reflect the official policy or position of their employers or any other agency or organization.

References

ABR, Inc., Sigma Plus, Statistical Consulting Services, Stephen R. Braund & Associates, Kuukpik Subsistence Oversight Panel, Inc., 2007. Variation in the Abundance of Arctic Cisco in the Colville River: Analysis of Existing Data and Local Knowledge, Volumes I and II. Prepared for the U.S. Department of the Interior, Minerals Management Service Alaska OCS Region, Anchorage, Alaska. Technical Report No. MMS 2007-042. https://www.boem.gov/sites/default/files/boem-newsroom/Library/Publications/2007/2007_042.pdf.

Aikenhead, G.S., Ogawa, M., 2007. Indigenous knowledge and science revisited. Cult. Stud. Sci. Educ. 2 (3), 539–620.

Briscoe, M.G., 2008. Collaboration in the ocean sciences: best practices and common pitfalls. Oceanography 21 (3), 58–65.

Coleman, D., Battiste, M., Henderson, S., Findlay, I.M., Findlay, L., 2012. Different knowings and the Indigenous humanities. ESC Engl. Stud. Can. 38 (1), 141–159.

Freire, P., 1968. Pedagogy of the Oppressed. Seabury Press, New York.

Galginaitis, M., 2014a. Monitoring Cross Island Whaling Activities, Beaufort Sea, Alaska: 2008–2012 Final Report, Incorporating ANIMIDA and cANIMIDA (2001–2007). OCS Study BOEM 2013-212. U.S. Department of the Interior, Bureau of Ocean Energy Management, Alaska OCS Region, Anchorage, Alaska, p. 208. https://www.arlis.org/docs/vol1/BOEM/CrossIsland/FinalReport2008-12/FinalReport2012.pdf.

Galginaitis, M., 2014b. An overview of cross island subsistence bowhead whaling, Beaufort sea, Alaska. Alaska J. Anthropol. 12 (1).

Gryba, R., Huntington, H.P., Von Duyke, A.L., Adams, B., Frantz, B., Gatten, J., Harcharek, Q., Olemaun, H., Sarren, R., Skin, J., Henry, G.H., 2021. Indigenous knowledge of bearded seal (*Erignathus barbatus*), ringed seal (*Pusa hispida*), and spotted seal (*Phoca largha*) behaviour and habitat use near Utqiaġvik, Alaska. Arc. Sci. https://doi.org/10.1139/as-2020-0052.

Kendall Jr., J.J., Brooks, J.J., Campbell, C., Wedemeyer, K.L., Coon, C.C., Warren, S.E., Auad, G., Thurston, D.K., Cluck, R.E., Mann, F.E., Randall, S.A., 2017. Use of traditional knowledge by the United States Bureau of Ocean Energy Management to support resource management. Czech Polar Rep. 7 (2), 151–163.

Latour, B., Woolgar, S., 1979. Laboratory Life. The Social Construction of Scientific Facts. Sage Publications, Beverly Hills, Calif., and London.

National Oceanographic Partnership Program (NOPP), 1997. Public Law 104-210, Title II, Subtitle E, Sec. 281-282. 10 U.S. Code 7901-7903. 1997. Also Reauthorized as Public Law 116-283, Title II, Subtitle E Sec. 1055, 2021.

Smith, L.T., 1999. Decolonizing Methodologies: Research and Indigenous Peoples, second ed. Zed Books, London and New York. 2012.

Whyte, K.P., Brewer, J.P., Johnson, J.T., 2016. Weaving indigenous science, protocols and sustainability science. Sustain. Sci. 11 (1), 25–32.

CHAPTER 12

Global marine biodiversity partnership

Francis K. Wiese[a] and Guillermo Auad[b]
[a]Stantec Consulting Services, Inc., Anchorage, AK, United States; [b]Office of Policy and Analysis, Bureau of Safety and Environmental Enforcement, U.S. Department of the Interior, Sterling, VA, United States

Preface

Today is September 5, 2051, and we are writing a synthesis article evaluating past efforts to rebuild marine life in the Global Ocean while seeking inspiration from the vast and beautiful ocean view seen from the Yucatan peninsula. We used to scuba dive here in our youth. Earlier this century, much was written about what the world may look like around 2050; the effects of climate change, the consequences of a world with a 23% population increase from 2020 levels, concerns about food security, hopes of an energy transition toward net-zero carbon emissions, and goals for ecosystem restoration and marine protected areas. Sitting here together, both past 80 years old, we reflect on how an extraordinary group of people came together to help rebuild marine life and realize the world we live in today. But we should start at the beginning of this journey, more than thirty years ago.

Introduction

January 2021: the beginning of the UN Decade on Ecosystem Restoration and the Decade of Ocean Science for Sustainable Development — it would prove to be a turning point for marine protection, restoration, and sustainable use, that would help restore marine biodiversity, habitats, and ecosystem structure, function, and services.

By the end of 2020, 90% of economically important species had declined, more than half of seagrass and wetland habitat was degraded, water quality had declined, and the rate of species invasions was increasing (Lotze et al., 2006). No marine area remained unaffected, and nearly half were impacted by multiple stressors, primarily habitat degradation, exploitation, and climate change (Halpern et al., 2008). Whereas all three stressors could act directly and independently, climate change exacerbated and accelerated impacts through the effects of ocean warming, sea-level rise, changes in ocean circulation and water column properties, pollutants, and reduction in habitat and population resilience (Worm and Lotze, 2021). In turn, reduced biodiversity compromised the adaptive capacity and diversity responses to said stressors, destabilizing ecosystem structure and function, threatening the oceans ability to provide ecosystem services, and thus securing the economies and livelihoods that depended on them (Schindler et al., 2010).

It was already known at the time that biodiversity was responsible for a plethora of ecosystem functions and services (Gamfeldt et al., 2015) and critical for human health and wellbeing (Diaz et al., 2006). In 2010 and according to the Organisation for Economic Co-operation and Development (OECD), the ocean supported a global ocean economy valued at US$ 1.5 trillion, primarily from fisheries, maritime transport, and tourism, and with 3.5 billion people dependent on the ocean for their protein. By 2020 the global degradation of estuarine, coastal, and marine ecosystems, had decreased the number of viable fisheries by a third, the provision of nursery habitats such as oyster reefs, seagrass beds, and wetlands by two-thirds, and water filtering and detoxification services provided by suspension feeders, submerged vegetation, and wetlands, by nearly two-thirds (Barbier et al., 2011). Additionally, while 40% of the world's population lived within 50 miles of the coast, including two-thirds of the world's largest cities, the loss of coastal vegetation decreased coastal protection from flooding and storms (Koch et al., 2009). Despite all this knowledge, the Marine Conservation Institute and its Atlas of Marine Protection estimated that in 2020, only about five percent of the ocean was being managed through the use of true Marine Protected Areas (Briggs, 2020), defined as areas where no destructive activities and, at most, light extractive activities, were allowed.

Fertile Ground

During the 2020s the timing was right to form a large, multi-disciplinary collaboration, that would co-design and co-produce an approach to rebuild global marine biodiversty by integrating climate change and pollution mitigation with the protection of habitats and biodiversity, and with sustainable harvests and resource extractions. The need was clear. We needed to restore and protect the diversity of marine species and habitats, and secure a sustainable use of the oceans and the stability of ecosystem services needed for the survival and well-being of people around the globe.

That of course was easier said than done, recognizing that biodiversity—a complex manifestation of ecological processes, evolutionary adaptations, and the severity and frequency of anthropogenic stressors—is not easily restored, and knowing that the reliability of ecosystem services erodes faster than indicated by species or habitat loss alone (Schindler et al., 2010). The need for a comprehensive approach to focus on the socio-economic-ecological marine system emerged. While we were successful in individual aspects of such endeavors, e.g. global population recovery of humpback whales, restoration of mangroves in the Mekong Delta, sustainable coastal fisheries in the Philippines, the challenge was to integrate the different temporal and spatial scales of the multiple components of the system, scaling these solutions up to where they could make a meaningful contribution within a few decades, and addressing the issues of national and international governance (e.g. van Tatenhove et al., 2021).

Although not integrated yet, several efforts to address these issues had already come to fruition at various scales in the areas of policy, management, conservation, and research, and had led to a decrease of species threatened with global extinction from 18% in 2000 to 11.4% in 2019 (Duarte et al., 2020).

Policy: In addition to the continued support of existing international conventions for the protection of species and habitats, several new international policies and established goals were adopted in 2021 under the Global Biodiversity Framework of the Convention on Biological Diversity (CBD, 2021) and supported by the UN Sustainable Development Goal 14 (SDG 14: Life below Water), addressing ocean stewardship, conservation, and sustainable use.

Adding more momentum in those days was the new Treaty of the High Seas under the UN Convention for the Law of the Sea (UNCLOS) which called for the conservation and sustainable development of marine biological diversity in areas beyond national jurisdiction, passed with government co-sponsors from every voting nation. This action sent a clear message to support this neglected half of our planet (High Seas Alliance). Importantly, the treaty included specific measures to ensure that environmental impact assessments would be consistent, comprehensive, accountable, and rigorous. The treaty also established a global decision-making body.

Many leaders stepped up during Conference of the Parties and Convention on Biological Diversity (CBD) meetings, and agreed on a coherent policy framework across the nexus of biodiversity, health, food, water, and climate change (Turney et al., 2020). Among other things, this included agreements by participating countries to Nationally Determined Contributions toward rebuilding marine life, just like they had recently done for mitigating climate change under the 2015 Paris Climate Agreement.

There was an additional effort during the early and mid 2020s to examine and improve research policies to promote partnerships (Chapter 11, Research Partnerships and Policies: A Dynamic and Evolving Nexus). By the time the GMBP got implemented in the mid 2020s, these were well in place at multiple levels of government across the globe and within national private and academic institutions, allowing the partnerships' *"form to follow function,"* like our dear colleague Lauren Alexander Augustine from the National Academies always said, rather than being constrained by bureaucratic idiosyncrasies.

Management: Ocean resources management underwent a national and international transformation during the UN Decade of Ocean Science for Sustainable Development. At an international level, the new High Seas Treaty benefited from excellent work by the Partnership for Regional Ocean Governance (PROG, Wright et al., 2017), and resulted in a new network of coordinated Regional Ocean Management Organizations (ROMOs), bringing together nations and stakeholders at an ecosystem scale and working together across sectors and national boundaries to manage shared resources in adjacent High Seas areas.

A Global Coastal Zone Management Network was also established. Through the early 2000s, an increasing number of studies suggested that new ways of governing coastal zones in a manner that protected ecosystem, community, and individual well-being were needed to reverse current trends (Vodden, 2015).

From a fisheries management perspective, there was a move toward increased scientific assessments and community empowerment, recognizing that most fisheries, particularly in developing countries, were not being assessed and in poor condition (Worm and Lotze, 2021). Moreover, granting fishing communities clear, exclusive rights to their fisheries in certain areas was leading to increased sustainability (Stuchtey et al., 2020).

Conservation: During the 2016 IUCN World Conservation Congress, world leaders were urged to protect 30% of the oceans by 2030 (IUCN, 2017), reaffirming the definition of an MPA as "a clearly defined geographical space, recognized, dedicated and managed, through legal or other effective means, to achieve the long-term conservation of nature with associated ecosystem services and cultural values." Based on strong recommendations from several NGOs, particularly PEW (e.g., Briggs et al., 2018), the backing from marine scientists, and the awareness of leaders around the world, this target was not only adopted during the 2020s, but clearer categories were defined. These included, "fully" and "strongly protected" MPAs, and the concepts of "low-level non-industrial natural resource use," "industrial fishing," "artisanal fishing," "indigenous fishing" were clarified. Critically, world leaders agreed to more stringent management, monitoring, and enforcement, including submitting plans for such actions alongside MPA designations and made publicly available by the IUCN to promote effectiveness.

Research: The UN Decade of Ocean Science for Sustainable Development (Ocean Decade for short) was a key catalyst in strengthening the international cooperation needed to develop the research and innovative technologies to connect science with the needs of society. It spurred substantial initial investment, especially once the 2017 Global Ocean Science Report documented that despite the ocean's key role in food security, energy security, the global economy, and human well-being, worldwide government ocean research and development expenditures only accounted for 0.04%—4% of total research and development expenditures worldwide (IOC-UNESCO, 2017). The report also highlighted major disparities in the worldwide capacity to undertake research — the Ocean Decade had set out to address this key issue.

Also important was the 2019 Intergovernmental Science-Policy Platform on Biodiversity and Ecosystem Services (IPBES) decision to launch an extensive research program to explore the impact of climate change on biodiversity, including adaptive capacities, resilience, and the risk to ecosystem services (Pörtner et al., 2021). They would prove to be a key partner in our Global Marine Biodiversity Partnership (GMBP).

The Partnership

All these accomplishments in the areas of policy, management, conservation, and research led to further discussions and realizations linking food security and marine biodiversity for most areas of the Global Ocean. The need to deliver a shared, clear, and actionable vision gained significant momentum and traction across global forums. Below we summarize this time period spanning almost 5 years (2023—28) building the partnership and integrating and adapting to new ideas as it grew.

In setting up the GMBP, we had reviewed the relevant regional, national, and sub-national ongoing marine biodiversity efforts and considered the lessons learned from past partnership programs (Chapter 10, Lessons Learned from Nine Partnerships in Marine Research). It was clear that this partnership needed to be aligned around an evidence-based action plan with quantitative targets and metrics for success; would require science, business, and educational plans; and be supported by existing policy frameworks (Duarte et al., 2020). Several accomplishments resonated constructively: (a) the effective collaborative approach of the Arctic Council in particular, and of the Arctic community in general, in delivering legally binding agreements on environmental issues, (b) the proven effectiveness of the Belmont Forum (Chapter 2, The Belmont Forum) as a partnership of partnerships in delivering numerous international projects from start to end, (c) the many scientific outcomes of the Malaspina Expedition on marine biodiversity in the Global Ocean (Duarte, 2015), and (d) the then recent and very well-received reports on Arctic biodiversity produced under the Arctic Council's working group on Conservation of Arctic Flora and Fauna (CAFF, e.g., Meltofte et al., 2013).

The Vision

To promote a culture of care (Francis, 2015; Klein and Bayne, 2007), and actively restore the integrity of marine ecosystem structure, function, and resilience around the World Ocean, in tandem with mitigating the impacts of climate change and providing sustainable resources for future generations.

The Partners

The Arctic Council's Sustainable Development Working Group recognized existing efforts and the value of bridging organizations (Muller-Karger et al., 2021) to advance knowledge and actions (Duarte et al., 2020) and called for a global effort to both integrate and expand said knowledge into actionable deliverables. Building on its successful international partnerships on Arctic Resilience and Ocean Sustainability, the Belmont Forum reached out to its existing members, and new ones quickly joined based on the pragmatic need to make decisions on food security rooted in the goal of rebuilding marine life. Regional champions were identified to help bring key people and resources to the table, and partnerships were cemented through memorandums of understanding wherein all parties specified their roles and responsibilities as well as self-imposed metrics to create transparency and accountability.

After 3 years of work, the GMBP included Indigenous Peoples and local community organizations from 23 countries across six continents; representatives from private industry including the transportation, energy, mining, tourism, and seafood sectors; global and national NGOs and citizen science organizations; natural and social scientists from 43 universities; the Natural Capital Finance Alliance; the High Seas Alliance; UNEP-WCMC and their network through the Regional Sea Program,particularly those covering the Caribbean Region, East Asian Seas, Eastern Africa Region, Mediterranean Region, North-West Pacific Region, and Western Africa Region; the Global Island Partnership; the newly established Regional Ocean Management Organizations and the Global Coastal Zone Management Network; IPBES; and representatives from government resource management agencies across all participating nations. More partners would follow as the GMBP evolved over the next 20+ years.

In the fall of 2026, the GMBP was born. Much still needed to be done, however, before we got underway, including co-producing science; business, and communication plans; developing a leadership model, and securing funding.

The Approach

Similar to that highlighted in the Ocean Decade, and as outlined in Kelly and Fisher (2021) for the Arctic, the basic principle behind the GMBP was one of co-production. Specifically, we used tools appropriate for advancing and sustaining partnerships, recognizing them as complex social systems (Chapter 10, Lessons Learned from Nine Partnerships in Marine Research; Chapter 11, Research Partnerships and Policies: A Dynamic and Evolving Nexus). This collaboration between knowledge producers, decision-makers, community memebers, and key stakeholders, resulted in the design and application of co-produced approaches to adapt to, mitigate, and anticipate environmental, social, cultural, and economic consequences of a changing ocean and decreasing biodiversity, and sharing the knowledge with key audiences in clear and timely ways. As noted by Duarte et al. (2020), rebuilding marine life would also require taking into account the different states of readiness and capacity to implement actions across different nations, with specific considerations to remain flexible to adapt to variable cultural settings and locally designed approaches.

Topically, we decided to focus on the following inter-dependent priorities: *protection, restoration, mortality reduction*, and *blue economy*, considering climate change in all aspects as a stressor and as a mitigation target.

Protection focused primarily on MPAs. By the mid 2020s, there were over 15,000 MPAs worldwide covering 27.4 million km^2 (7.6% of the global ocean), with the twenty largest comprising about 70% of existing global coverage. MPAs can safeguard ocean areas from destructive and/or extractive activities such as unsustainable fishing, and yield significant benefits to ecosystems and the people that depend on them. They can also promote biological processes that build resilience against changing environmental conditions; and when containing reefs or aquatic vegetation, can enhance and maintain carbon sequestration and storage, thus helping to mitigate the effects of climate change (Roberts et al., 2017). Their effectiveness, however, varies substantially, with key characteristics including high levels of protection and enforcement, clearly defined boundaries, stakeholder engagement, and community involvement, sustained resources, and large size (Briggs et al., 2018).

The partners decided to focus on examining the effectiveness of current MPAs using specific biodiversity metrics, detailing socio-economic benefits analysis, establishing additional national, transboundary, and high seas protected areas focusing on key species, habitats, ecosystem services to help meet the 30% global ocean conservation target; and helping in the design and implementation of more stringent management, monitoring, and enforcement actions.

Restoration focused on active ecosystem restoration of primarily coastal marine habitats in line with goals set forth in the UN Decade on Ecosystem Restoration and following up on the components of the Marine Ecosystem Restoration in Changing European Seas (MERCES) program that had ended in 2020. Functional coastal ecosystems support a wide array of services including water quality, biodiversity, food security, tourism, protection from flooding, storms, and coastal erosion, and carbon sequestration and storage.

Back then, implementing a coastal and marine restoration program, however, still had its challenges, e.g. see Danovaro et al., 2021; van Tatenhove et al., 2021. As a result, once the team had prioritized habitats and locations for restoration and established clear and consistent monitoring and success metrics, all through extensive local stakeholder engagement, the early phases of the program focused extensively on local capacity building, integrating existing projects, understanding local and global stressors, feedbacks and functional thresholds, identifying appropriate donor sites where possible, developing culture or grow-out methods where that was not possible, e.g. corals resistant to temperatures and acidity levels expected by the end of the 21st century (Duarte et al., 2020), and understanding and optimizing species interactions that expedite recovery. In addition, we included modules on offshore habitat restoration, primarily focused on habitat improvements for fish, seabirds, sea turtles, and marine mammals through a reduction in by-catch, plastic pollution, and anthropogenic underwater noise, which linked to our *mortality reduction* topic.

Key to this success were additional considerations of permanency, where each site was modeled into the future based on regional climate change projections to ensure restored habitats would be resilient; initial and recurring socio-economic benefits analysis; natural capital valuation; engagement with government and stakeholders to uphold long-term support (and show the link to achieving blue-economy goals, as noted below); and linking these restoration activities, where appropriate, to increased carbon sequestration, providing carbon offsets to support net-zero goals by industries and governments. At the same time, our partnerships with urban planners and community leaders concerned with sea-level rise and overall coastal resilience, allowed for the increased emphasis on nature-based-solutions, thus connecting coastal ecosystem restoration with coastal protection, while also helping to mitigate climate change (Narayan et al., 2016; Griscom et al., 2017; Seddon et al., 2020; Von Unger et al., 2020).

Mortality Reduction focused on reducing direct and indirect mortality of fish, seabirds, sea turtles, and marine mammals and rebuilding fish spawning stocks, while supporting efforts to increase food production. Fisheries effectively truncate the age and size structure of target fish by preferentially removing the larger and older individuals. As a result, the success of the fishery becomes more dependent on the recruitment of new individuals, a process strongly influenced by climate change and its associated variability (Worm and Lotze, 2021). Reducing fishing mortality and rebuilding spawning stocks then was the a feasible means of mitigating the impacts of climate change on recruitment.

Supported by the offshore restoration efforts detailed above and in close collaboration with stakeholders and regulatory agencies in the fishing industry, the main thrust was to increase and support the number of fish stocks with science-based assessments, increase the inclusion of environmental variables into these assessments, increase use of management scenario evaluations driven by policy options and climate change scenarios, adopt more adaptive management strategies, and support flexible institutions and social networks in multi-level governance systems (Hughes et al., 2005; Auad et al., 2018), with more explicit accounting of animal movements and considerations of dynamic marine spatial management (Sequeira et al., 2019), and cumulative impacts of other stressors (Worm and Lotze, 2021). These approaches needed to be implemented at all levels, and whereas much work was done directly with the resource managers in the different jurisdictions, a lot of time and effort was invested in empowering communities and modifying incentives to align economic and conservation outcomes (Hughes et al., 2005; Stuchtey et al., 2020). This turned out to be extremely effective and led those communities to embrace sustainable practices increasing food and financial security, and improving their well-being.

The blue economy was focused on aligning economic and biodiversity outcomes across strategically selected regions. If time and resources were spent on ecosystem restoration, protecting and increasing the number and coverage of MPAs, and reducing exploitation levels of marine resources, then economic benefits needed to be realized, and alternative economic opportunities needed to be supported. The blueprint for this, driven by providing solutions for climate change mitigation, had been laid out by the High-Level Ocean Panel in late 2019 (Hoegh-Guldberg et al., 2019). Concomitantly, sustainable economic processes were supported by consistent principles, practices, and policies in line with well-known regenerative dynamics of complex social systems (Goerner et al., 2009; Fath et al., 2019). It provided ocean-based climate change mitigation options, which at the same time could be used for diversification of coastal economies in the areas of ocean energy, aquaculture, and seaweed farming. Combined with increased capacity building, increased tourism supported by ecosystem restoration efforts and marine protected areas, and more sustainable locally managed fisheries, this focus resulted in ongoing increases in human well-being, food, education, and economic opportunities in our focus areas.

The Leadership

Once the research focus and approaches were agreed on (*function*), our leadership model (*form*) developed almost naturally, with clear champions emerging across topics and geographies during the many months of in-person and virtual meetings. The leadership model followed the program structure in that leadership champions came from all levels and backgrounds. At the highest level, we decided on topical leadership teams that would serve as focal points for each of the four focus areas, plus an executive director, and chief climate change, sustainability, and communication officers. Country coordinators for each of participating countries were appointed, and together, this entire team would become known as the Global Marine Biodiversity Collaborative. The role of the Collaborative was to function as a steering committee that would realize and sustain the vision over the two to three decades required to rebuild marine biodiversity while supporting a blue economy. They were further aided by an Advisory Board made up of community members, public and private decision-makers, and natural and social scientists.

Operational Phase
Funding
Initial funding to establish the GMBP was key. It came in part from the UN, the World Bank, the EU Framework Initiatives, and several country-specific science foundations, who all recognized that an effort at this scale was requiered, but that a comprehensive planning effort needed to first be supported.

Once the GMBP was in place, however, substantial long-term funding had to be secured. Duarte et al. (2020) had estimated it might take two to three decades to rebuild marine life to 90% of undisturbed baselines, recognizing and that at least US$10—20 billion per year would be required to extend protective actions to 50% of the ocean space, plus substantial additional funds for restoration. Although garnering such a level of support over such an extended time-period seemed daunting, we benefited from several previous economic analyses and funding decisions of the early 2020s. A detailed business plan was developed that illustrated an economic return on investment (ROI) of at least 10:1 and the creation of over one million jobs (Barbier et al., 2018). This ROI would result from ecotourism, the rebuilding and redevelopment of the global seafood and aquaculture industry, the energy sector, and government savings through reduction of damage to coastal areas from sea-level rise and extreme weather events, long-term savings due to climate change mitigation effects, reduction in costs associated with climate refugees, and other associated local economic diversification.

With the economic case made and the backing of the Organization for Economic Co-operation and Development (OECD Ocean) and the World Economic Forum, who had recognized the immediate need to mitigate the impacts from habitat degradation and climate change, we were able to tap into a series of international and national funding sources. Internationally, the UN's Green Climate Fund's (https://www.greenclimate.fund/) goal to mobilize $100 billion annually to assist developing countries to adapt to climate change had become reality in 2024 after the financial clauses of Article 6 of the Paris Agreement finally fell into place. As suggested several years before by Duarte et al. (2020), a portion of this got allocated for marine conservation, restoration, and nature-based solutions, with a substantial portion allocated to our program. The Belmont Forum also played a fundamental role in international funding coordination, allowing for co-ordinated implementation (Chapter 2, Belmont Forum Partnerships).

Coordination and Communication

Annual meetings, including information transfer meetings within and across national topical action teams, quarterly collaboration, and monthly leadership calls, all helped to keep things on track and coordinated. Internal coordination was further aided by a project and data portal which – in addition to all the data and metadata upload, storage, and visualization capabilities – also allowed both for a project-wide chat forum where anyone could reach out to anyone else worldwide and ask questions and share findings and experiences.

We also found ourselves in the *"once you build it, they will come"* situation whereby once the GMBP was on its feet, many other ongoing efforts reached out to us for even broader coordination and integration. An important component of this were ongoing citizen science initiatives on marine biodiversity, climate change effects, and invasive species tracking, e.g. Mannino and Balistreri (2018); Pecl et al. (2019), that had proven very effective in ocean monitoring and informing resource management agencies (Chapter 6, Partnering with the Public: The Coastal Observation and Seabird Survey Team). The GMBP provided a means for these efforts to expand worldwide, further supporting public engagement, education, and capacity building.

Challenges

The GMBP was not without its challenges. Realizing the benefits from restoring marine biodiversity took time, particularly as the requisite initial steps involved first reversing the negative trends. This time lag in widely recognized results made sustained public and financial support difficult during the mid to late 2030s. The Collaborative fell into a crisis struggling to maintain the cohesiveness of the partners across the globe. Additional effort was needed to convince new governments and funding organization officials that had not been there a decade earlier during the program's inception and visioning, that continued investment was warranted. Likewise, public support needed to be maintained.

But lessons learned from past marine research partnerships had been studied (Chapter 10, Lessons Learned from Nine Partnerships in Marine Research), and so out of extensive internal and external communication and collaborations efforts, our robust, yet flexible and diverse leadership, team, and funding models built into the program design from day one, were there to revitalize and sustain the efforts needed to achieve the goals and vision of the program.

Epilogue

The effects of climate change and human activities have differed substantially across the Global Ocean and on the different physical, chemical, and biological aspects of the marine ecosystem. Climate change, particularly ocean warming, has caused substantial changes in species distributions as well as in adaptive capabilities of some taxa, especially those with faster generation times and exposed to more extremes, such as bacteria, plankton, and corals. But just as global collaboration managed to avert the worst of the climate crisis and the collapse of the Arctic by 2050 (Wiese, 2021) by implementing an international adaptive strategy (Auad and Fath, 2021), the efforts of the GMBP and many other local and international initiatives managed to first halt, and then reverse, the decrease in marine species and habitat biodiversity during this same timeframe.

Once reliable scientific information was made available, several governance-relevant aspects were addressed, which helped transform knowledge into actionable and effective decisions. Whereas restoration activities and monitoring continued throughout the duration of the GMBP, efforts were needed to further institutionalize active restoration with overarching restoration goal setting and monitoring in mind, particularly at the regional level (van Tatenhove et al., 2021). Looking back now from 2051, a fascinating outcome was that initially no-legally binding agreements were needed to produce governmental decisions to restore marine life. Because a significant portion of the overall funding came from government agencies of many nations, and due to the wealth of information produced through the inclusion of scientists and knowledge holders from all participating nations, many governments first acted regionally to protect their resources and increase food security. Encouraged by the successes achieved in regional seas by these nations, three more outcomes resulted: (1) other nations followed similar paths, (2) legally binding agreements were implemented in international waters, and (3) policies were created and amended, during and after the UN Decade of Ocean Science (2021–30), to further facilitate international partnerships as advocated by many since the start of this century, e.g., Duda (2006), Claudet et al. (2020), Pendleton et al. (2020).

Climate velocity, a term first coined by Loarie et al. (2009) and further developed by Molinos et al. (2016) in terms of the speed of spatial redistribution of marine biodiversity, slowed down. Some species disappeared from some regions, but through the combined efforts of climate mitigation, MPAs, and habitat restoration, have been slowly returning. Marine protected areas now cover 32% of the worlds' oceans (a work in progress) and adaptive management efforts that are fully integrated into climate scenarios and changes in key environmental variables are now the norm. Over 90% of exploited fish stocks are scientifically assessed, designations of overexploited stocks decreased by 83%, and the IUCN Red List of marine species threatened with global extinction decreased 3.4%.

Today, in 2051, the world population has reached 10 billion, and the value-added generated by the ocean-based industry globally exceeds U$ 5 trillion. As a consequence of sustainable ocean management, and as predicted by Stuchtey et al. (2020) three decades ago, the ocean now produces almost six times more food than in 2020, generates 40 times more renewable energy, has contributed one-fifth of the reductions in greenhouse gases in the atmosphere, has helped to lift millions of people out of poverty, and economic and environmental resilience has increased.

Ultimately, the prediction that restoring marine species and habitat biodiversity would lead to an increase in productivity and resilience and help mitigate climate change and other disturbances on marine systems (Duarte et al., 2020; Worm and Lotze, 2021), proved true. Continued global collaborations and partnerships are needed, and much remains to be done to ensure all these positive trends continue. It turns out that investments in ocean restoration, protection, and a sustainable ocean economy, were indeed not just good for the ocean, but has benefited everyone. It is wonderful to still be able to scuba dive and enjoy the pristine and productive waters off the Yucatan Peninsula as we teach our grandchildren here with us, to be appreciative of the oceans' beauty and role in nature. This state of the oceans we are enjoying today was not certain thirty years ago.

Acknowledgments

The authors are very grateful to Drs. James J. Kendall, Jr., Brendan P. Kelly, and Melbourne G. Briscoe for their review of this chapter. They have enhanced the quality and clarity of its content. The views and opinions expressed in this chapter are those of the authors and do not necessarily reflect the official policy or position of their employers or any other agency or organization.

References

Auad, G., Fath, B.D., 2021. Recipes for a flourishing Arctic. In: Wassmann, P. (Ed.), Whither the Arctic Ocean? Research, Knowledge Needs and Development en Route to the New Arctic. Fundación BBVA, Bilbao, Spain, pp. 75—86.

Auad, G., Blythe, J., Coffman, K., Fath, B.D., 2018. A dynamic management framework for socio-ecological system stewardship: a case study for the United States Bureau of Ocean Energy Management. J. Environ. Manag. 225, 32—45.

Barbier, E.B., Hacker, S.D., Kennedy, C., Koch, E.W., Stier, A.C., Silliman, B.R., 2011. The value of estuarine and coastal ecosystem services. Ecol. Monogr. 81 (2), 169—193.

Barbier, E.B., Burgess, J.C., Dean, T.J., 2018. How to pay for saving biodiversity. Science 360 (6388), 486—488.

Briggs, J., 2020. How Much of the Ocean Is Really Protected? the Pew Bertarelli Ocean Legacy Project. https://www.pewtrusts.org/-/media/assets/2020/06/pbol-updates/how-much-of-the-ocean-is-protected.pdf.

Briggs, J., Baez, S.K., Dawson, T., Golder, B., O'Leary, B.C., Petit, J., Roberts, C., Rogers, A., Villagomez, A., 2018. Recommendations to IUCN to Improve Marine Protected Area Classification and Reporting. The Pew Bertarelli Ocean Legacy Project. https://www.pewtrusts.org/-/media/assets/2018/02/recommendations-to-iucn-on-implementing-mpa-categories-for-printing.pdf.

Claudet, J., Bopp, L., Cheung, W.W., Devillers, R., Escobar-Briones, E., Haugan, P., Heymans, J.J., Masson-Delmotte, V., Matz-Lück, N., Miloslavich, P., Mullineaux, L., 2020. A roadmap for using the UN decade of ocean science for sustainable development in support of science, policy, and action. One Earth 2 (1), 34—42.

Convention on Biological Diversity (CBD), 2021. First Draft of the Post-2020 Global Biodiversity Framework. UN Environmental Programme CBD/WG2020/3/3. https://www.cbd.int/doc/c/abb5/591f/2e46096d3f0330b08ce87a45/wg2020-03-03-en.pdf.

Danovaro, R., Aronson, J., Cimino, R., Gambi, C., Snelgrove, P.V., Van Dover, C., 2021. Marine ecosystem restoration in a changing ocean. Restor. Ecol.

Díaz, S., Fargione, J., Chapin III, F.S., Tilman, D., 2006. Biodiversity loss threatens human well-being. PLoS Biol. 4 (8), e277.

Duarte, C.M., 2015. Seafaring in the 21st century: the Malaspina 2010 circumnavigation expedition. Limnol. Oceanogr. Bull. 24 (1), 11—14.

Duarte, C.M., Agusti, S., Barbier, E., Britten, G.L., Castilla, J.C., Gattuso, J.P., Fulweiler, R.W., Hughes, T.P., Knowlton, N., Lovelock, C.E., Lotze, H.K., 2020. Rebuilding marine life. Nature 580 (7801), 39—51.

Duda, A.M., 2006. Policy, legal and institutional reforms for public—private partnerships needed to sustain large marine ecosystems of East Asia. Ocean Coast Manag. 49 (9—10), 649—661.

Fath, B.D., Fiscus, D.A., Goerner, S.J., Berea, A., Ulanowicz, R.E., 2019. Measuring regenerative economics: 10 principles and measures undergirding systemic economic health. Glob. Trans. 1, 15—27.

Francis, P, 2015. Laudato si: On care for our common home. Our Sunday Visitor Publishing, Huntington, Indiana, USA.

Gamfeldt, L., Lefcheck, J.S., Byrnes, J.E., Cardinale, B.J., Duffy, J.E., Griffin, J.N., 2015. Marine biodiversity and ecosystem functioning: what's known and what's next? Oikos 124 (3), 252–265.

Goerner, S.J., Lietaer, B., Ulanowicz, R.E., 2009. Quantifying economic sustainability: implications for free-enterprise theory, policy and practice. Ecol. Econ. 69 (1), 76–81.

Griscom, B.W., Adams, J., Ellis, P.W., Houghton, R.A., Lomax, G., Miteva, D.A., Schlesinger, W.H., Shoch, D., Siikamäki, J.V., Smith, P., Woodbury, P., 2017. Natural climate solutions. Proc. Natl. Acad. Sci. U.S.A. 114 (44), 11645–11650.

Halpern, B.S., Walbridge, S., Selkoe, K.A., Kappel, C.V., Micheli, F., D'Agrosa, C., Bruno, J.F., Casey, K.S., Ebert, C., Fox, H.E., Fujita, R., 2008. A global map of human impact on marine ecosystems. Science 319 (5865), 948–952.

Hoegh-Guldberg, O., Lovelock, C., Caldeira, K., Howard, J., Chopin, T., Gaines, S., 2019. The Ocean as a Solution to Climate Change: Five Opportunities for Action. Report. World Resources Institute, Washington, DC. Available online at: http://www.oceanpanel.org/climate.

Hughes, T.P., Bellwood, D.R., Folke, C., Steneck, R.S., Wilson, J., 2005. New paradigms for supporting the resilience of marine ecosystems. Trends Ecol. Evol. 20 (7), 380–386.

IOC-UNESCO, 2017. In: Valdés, L., et al. (Eds.), Global Ocean Science Report - the Current Status of Ocean Science Around the World. UNESCO, Paris. https://en.unesco.org/gosr.

IUCN, 2017. A Summary Report of the 2016 International Union for Conservation of Nature (IUCN) World Conservation Congress. https://enb.iisd.org/events/2016-international-union-conservation-nature-iucn-world-conservation-congress planet/summary?utm_campaign=RSS2.0&utm_content=2016–09 13&utm_medium=feed&utm_source=www.iisd.ca.

Kelly, B.P., Fisher, A.M., 2021. Complex collaboration tools for a sustainable arctic. In: Wassmann, P. (Ed.), Whither the Arctic Ocean? Research, Knowledge Needs and Development en Route to the New Arctic, vol. 3. Fundación BBVA, Bilbao, Spain, p. p260.

Klein, H.J., Bayne, K.A., 2007. Establishing a culture of care, conscience, and responsibility: Addressing the improvement of scientific discovery and animal welfare through science-based performance standards. ILAR Journal 48 (1), 3–11.

Koch, E.W., Barbier, E.B., Silliman, B.R., Reed, D.J., Perillo, G.M., Hacker, S.D., Granek, E.F., Primavera, J.H., Muthiga, N., Polasky, S., Halpern, B.S., 2009. Non-linearity in ecosystem services: temporal and spatial variability in coastal protection. Front. Ecol. Environ. 7 (1), 29–37.

Loarie, S.R., Duffy, P.B., Hamilton, H., Asner, G.P., Field, C.B., Ackerly, D.D., 2009. The velocity of climate change. Nature 462 (7276), 1052–1055.

Lotze, H.K., Lenihan, H.S., Bourque, B.J., Bradbury, R.H., Cooke, R.G., Kay, M.C., Kidwell, S.M., Kirby, M.X., Peterson, C.H., Jackson, J.B., 2006. Depletion, degradation, and recovery potential of estuaries and coastal seas. Science 312 (5781), 1806–1809.

Mannino, A.M., Balistreri, P., 2018. Citizen science: a successful tool for monitoring invasive alien species (IAS) in Marine Protected Areas. The case study of the Egadi Islands MPA (Tyrrhenian Sea, Italy). Biodiversity 19 (1–2), 42–48.

Meltofte, H., Barry, T., Berteaux, D., Bültmann, H., Christiansen, J.S., Cook, J.A., Dahlberg, A., Daniëls, F.J., Ehrich, D., Fjeldså, J., Friðriksson, F., 2013. Arctic Biodiversity Assesment. Synthesis. Conservation of Arctic Flora and Fauna (CAFF).

Molinos, J.G., Halpern, B.S., Schoeman, D.S., Brown, C.J., Kiessling, W., Moore, P.J., Pandolfi, J.M., Poloczanska, E.S., Richardson, A.J., Burrows, M.T., 2016. Climate velocity and the future global redistribution of marine biodiversity. Nat. Clim. Change 6 (1), 83–88.

Muller-Karger, F., Kavanaugh, M., Iken, K., Montes, E., Chavez, F., Ruhl, H., Miller, R., Runge, J., Grebmeier, J., Cooper, L., Helmuth, B., 2021. Marine life 2030: forecasting changes to ocean biodiversity to inform decision-making: a critical role for the marine biodiversity observation network (MBON). Mar. Technol. Soc. J. 55 (3), 84—85.

Narayan, S., Beck, M.W., Reguero, B.G., Losada, I.J., Van Wesenbeeck, B., Pontee, N., Sanchirico, J.N., Ingram, J.C., Lange, G.M., Burks-Copes, K.A., 2016. The effectiveness, costs and coastal protection benefits of natural and nature-based defences. PloS one. 11 (5), e0154735.

Pecl, G.T., Stuart-Smith, J., Walsh, P., Bray, D.J., Kusetic, M., Burgess, M., Frusher, S.D., Gledhill, D.C., George, O., Jackson, G., Keane, J., 2019. Redmap Australia: challenges and successes with a large-scale citizen science-based approach to ecological monitoring and community engagement on climate change. Front. Mar. Sci. 6, 349.

Pendleton, L., Evans, K., Visbeck, M., 2020. Opinion: we need a global movement to transform ocean science for a better world. Proc. Natl. Acad. Sci. USA. 117 (18), 9652—9655.

Pörtner, H.O., Scholes, R.J., Agard, J., Archer, E., Arneth, A., Bai, X., Barnes, D., Burrows, M., Chan, L., Cheung, W.L., Diamond, S., Donatti, C., Duarte, C., Eisenhauer, N., Foden, W., Gasalla, M.A., Handa, C., Hickler, T., Hoegh-Guldberg, O., Ichii, K., Jacob, U., Insarov, G., Kiessling, W., Leadley, P., Leemans, R., Levin, L., Lim, M., Maharaj, S., Managi, S., Marquet, P.A., McElwee, P., Midgley, G., Oberdorff, T., Obura, D., Osman, E., Pandit, R., Pascual, U., Pires, A.P.F., Popp, A., Reyes- García, V., Sankaran, M., Settele, J., Shin, Y.J., Sintayehu, D.W., Smith, P., Steiner, N., Strassburg, B., Sukumar, R., Trisos, C., Val, A.L., Wu, J., Aldrian, E., Parmesan, C., Pichs-Madruga, R., Roberts, D.C., Rogers, A.D., Díaz, S., Fischer, M., Hashimoto, S., Lavorel, S., Wu, N., Ngo, H.T., 2021. IPBES-IPCC co-sponsored workshop report on biodiversity and climate change. IPBES and IPCC. https://doi.org/10.5281/zenodo.4782538. https://ipbes.net/sites/default/files/2021 06/20210609_workshop_report_embargo_3pm_CEST_10_june_0.pdf.

Roberts, C.M., O'Leary, B.C., McCauley, D.J., Cury, P.M., Duarte, C.M., Lubchenco, J., Pauly, D., Sáenz-Arroyo, A., Sumaila, U.R., Wilson, R.W., Worm, B., 2017. Marine reserves can mitigate and promote adaptation to climate change. Proc. Natl. Acad. Sci. USA. 114 (24), 6167—6175.

Schindler, D.E., Hilborn, R., Chasco, B., Boatright, C.P., Quinn, T.P., Rogers, L.A., Webster, M.S., 2010. Population diversity and the portfolio effect in an exploited species. Nature 465 (7298), 609—612.

Seddon, N., Chausson, A., Berry, P., Girardin, C.A., Smith, A., Turner, B., 2020. Understanding the value and limits of nature-based solutions to climate change and other global challenges. Philos. Trans. R. Soc. B 375, 20190120, 1794.

Sequeira, A.M.M., Hays, G.C., Sims, D.W., Eguíluz, V.M., Rodríguez, J.P., Heupel, M.R., Harcourt, R., Calich, H., Queiroz, N., Costa, D.P., Fernández-Gracia, J., 2019. Overhauling ocean spatial planning to improve marine megafauna conservation. Front. Mar. Sci. 6, 639.

Stuchtey, M., Vincent, A., Merkl, A., Bucher, M., Haugan, P.M., Lubchenco, J., Pangestu, M.E., 2020. Ocean Solutions that Benefit People, Nature, and the Economy. World Resource Institute. https://oceanpanel.org/ocean-action/files/ocean-report-short-summary-eng.pdf.

Turney, C., Ausseil, A.G., Broadhurst, L., 2020. Urgent need for an integrated policy framework for biodiversity loss and climate change. Nat. Ecol. & Evol. 4 (8), 996-996.

van Tatenhove, J.P., Ramírez-Monsalve, P., Carballo-Cárdenas, E., Papadopoulou, N., Smith, C.J., Alferink, L., Ounanian, K., Long, R., 2021. The governance of marine restoration: insights from three cases in two European seas. Restor. Ecol. 29, e13288.

Vodden, K., 2015. Governing sustainable coastal development: the promise and challenge of collaborative governance in Canadian coastal watersheds. The Canadian Geographer/Le Géographe canadien 59 (2), 167—180.

Von Unger, M., Herr, D., Seneviratne, T., Castillo, G., 2020. Blue NbS in NDCs. A Booklet for Successful Implementation (GIZ 2020) Blue Nature-Based Solutions in Nationally Determined Contributions by BlueSolutions. https://www.icriforum.org/wp-content/uploads/2020/12/NbS_in_NDCs._A_Booklet_for_Successful_Implementation.pdf.

Wiese, F.K., 2021. Why did the arctic not collapse? In: Wassmann, P. (Ed.), Whither the Arctic Ocean? Research, Knowledge Needs and Development en Route to the New Arctic. Fundación BBVA, Bilbao, Spain, pp. 229—238.

Worm, B., Lotze, H.K., 2021. Marine biodiversity and climate change. In: Climate Change. Elsevier, pp. 445—464.

Wright, G., Schmidt, S., Rochette, J., Shackeroff, J., Unger, S., Waweru, Y., Müller, A., 2017. Partnering for a Sustainable Ocean: The Role of Regional Ocean Governance in Implementing SDG14. PROG: IDDRI, IASS, TMG & UN Environment.

Index

Note: 'Page numbers followed by "b" indicate boxes, those followed by "f" indicate figures and those followed by "t" indicate tables.'

A

Alaska Fishery Observer Program, 96
Alaska Ocean Observing System, 2
Arctic Research Policy Act, 72
Arctic Social Science Program (ASSP), 158−159
Argo Data Management Team (ADMT), 63
Argo program
  Argo Science Team (AST), 53−54
  assessment, 65
  Core Argo array, 66
  cost, 64
  data management system, 63−64
  deployment locations, 64
  global Argo array implementation
    Argo float production and technology, 60−61
    depth range, 60
    float lifetime, 60
    regional pilot arrays, 61
    satellite networks, 60
    spin-up phase, 61
  Global Data Assembly Centers (GDACs), 53
  International governance, 61−63
  multi-national partnership
    Argo Science Team (AST), 57−58
    National Oceanographic Partnership Program (NOPP), 58−59
  operational Argo floats, 55f
  research-quality delayed mode (DM) dataset, 64
  spatial density, 54f
  sustainability, 64−66
  vulnerabilities and strengths, 67
  World Ocean Circulation Experiment (WOCE)
    conductivity-temperature-depth (CTD) profile, 56
    Data Assembly Centers (DACs), 63
    elements, 55
    scientific motivation, 56−57
    systematic observations, 55
Argo Science Team (AST), 53−54
Atlantic Canyons
  awards, 151
  contract award, 148−149
  findings, 149−150
  implementation, 147−148
  trust-based partnership, 150
  video data, 151

B

Baltimore Canyon, 133
BeachCOMBERS, 96
Belmont Forum partnerships
  challenges, 17, 30−31
  funding organizations, 17
  Future Earth (FE), 17
  partnering, 27−29
  partnership implementation
    award management and evaluation, 25−27
    call launch and review, 22−25
    Collaborative Research Action (CRA), 23−24
    consortia proposals, 24−25
    multi-lateral funding partnerships, 23f
    proposal team development, 22−23
    webinars, 22−23
  partnership processes
    funding model, 19−20
    joint call design, 20−21
    scoping process, 18−19
    resourcing, 29−30

Belmont Forum partnerships (*Continued*)
  stakeholder participation, 31
  transdisciplinarity, 17
  transnational engagement, 18
Bering Sea Ecosystem Interagency Working Group (BIAW), 2
Bering Sea project
  biological resources, 1
  communication, 13
  comprehensive data management support strategy, 10
  data and data management, 13
  ecological and economic consequences, 1
  end-to-end modeling, 12–13
  GIS tool, 10
  implementation, 12
    cognizant program, 4
    focal species and linkages, 6–7
    IPCC model, 4–5
    Local and Traditional Knowledge groups, 5–6
    NSF solicitation, 4–5
    proposals sharing, 4
    technical reviews, 5
  National Science Foundation (NSF), 1–2
  North Pacific Research Board (NPRB), 2
  partnership, 2–4, 12
  program management, 12
    interdisciplinary sampling, 8–9
    Road Map, 9
    Science Advisory Board (SAB), 7–8
    structure, 7–8, 8f
  scientific framing, 11
  seasonal ice, 1
  sustainability, 10–11
  target audiences, 9
  trust and team-building, 13
Biological Resources Division (BRD), 135–136
Bureau of Land Management (BLM), 112
Bureau of Ocean Energy Management (BOEM), 117

C

California Current Integrated Ecosystem Assessment (CCIEA), 96
California Current Large Marine Ecosystem (CCLME), 99
Chemo III study
  awards, 142
  contract award, 139–141
  deep-water coral habitats, 137
  findings, 141
  NOAA OE contacts, 138
  NOPP meeting, 138–139
  profile development, 137–138
  USGS participation, 139
  WWII shipwreck study, 137–138
Chemosynthetic communities, 132
Citizen science, 87–88
Coastal and ocean optical monitoring (Citclops), 88
Coastal Observation and Seabird Survey Team (COASST)
  communication, 102–103
  data collection sites, 91f
  funding, 101
  innovations, 94
  leadership, 101–102
  listening, 103
  motivation, 90–92
  number of participants, 93f
  organizational evolution
    current era, 99–100
    initial years, 92–95
    middle years, 95–99
  quality assurance/quality control (QAQC) measures, 90
  science, 103–104
  success and sustainability, 102–104
Collaborative Research Action (CRA), 23–24, 24f
Communication, 175
  Coastal Observation and Seabird Survey Team (COASST), 102–103
  Multi-Agency Rocky Intertidal Network (MARINe), 120–121
Consensus-based governance process, 116–117

Contracting Officer Representatives (CORs), 77
Convention on International Trade in Endangered Species (CITESs), 201

D

Deep-water ecosystems, Gulf of Mexico and northwest Atlantic
  Bureau of Ocean Energy Management (BOEM)
    Atlantic Canyons, 147–151
    Chemo III study, 137–142
    Department of the Interior (DOI), 131–132
    Environmental Studies Program (ESP), 131–132
    Lophelia II, 142–147
    structure and partners, 134, 134f
    timing, 133
  chemosynthetic communities, 132
  deep-water corals, 132–133
  habitat types, 132
  Minerals Management Service (MMS), 132
  multiple scientific disciplines, 131
  National Oceanographic Partnership Program (NOPP), 136–137
  NOAA's Office of Ocean Exploration and Research (OER), 135
  submarine canyons, 133

E

Ecosystem-based Fisheries Management (EBFM), 14
Ecosystem Modeling Committee (EMC), 3–4
Environmental science, 87
Environmental Studies Program (ESP), 131–132
Exclusive Economic Zones (EEZs), 61–62
External communications, 175
Exxon Valdez oil spill event, 112

F

Funding
  Belmont Forum partnerships, 19–20
  coastal Observation and Seabird Survey Team (COASST), 101
  global marine biodiversity partnership, 209–210
  Multi-Agency Rocky Intertidal Network (MARINe), 123–124
  Nansen Legacy project, 42
Future Earth (FE), 17

G

Global Biodiversity Framework of the Convention on Biological Diversity, 201
Global Coastal Zone Management Network, 202
Global Data Assembly Centers (GDACs), 53
Global marine biodiversity partnership
  challenges, 210–211
  climate change, 211
  coordination and communication, 210
  economically important species, 199–200
  fertile ground
    conservation, 202–203
    management, 202
    policy, 201
    research, 203
  funding, 209–210
  global ocean economy, 200
  leadership, 208–209
  needs, 200–203
  partnership
    mortality reduction, 207
    protection, 206
    restoration, 206
Global Ocean Observing System (GOOS), 53–54

Global Sustainable Coastal Development
  Collaboration Team, 202
Group of Program Coordinators
  (GPC), 19

I

Indefinite Delivery Indefinite Quantity
  (IDIQ) government
  contract, 73
Indian Reorganization Act (IRA), 159
Institutional Review Board (IRB), 163
Integrated Ecosystem Research Program
  focused on the Bering Sea
  (BSIERP), 2
Interagency Rocky Intertidal Monitoring
  Workshop, 120
Internal communication, 175

K

Knowledge production process, 185

L

Large marine ecosystem (LME), 87
Leadership, 173—174, 208—209
  Coastal Observation and Seabird Survey
  Team (COASST), 101—102
Lophelia II study
  awards, 147
  contract award, 143—144
  educational outreach, 145—146
  partnering, 146
  project implementation, 142—143
  scientific and management goals, 145

M

Mainstream science, 87
Marine Arctic Ecosystem Study
  (MARES)
  communication, 82
  creation process, 84
  formalized agreements, 80—81
  lead agency and research partner,
    79—80, 80f
  modular approach, 81
  partners, sectors and roles, 76t
  physical, chemical, and biological
    variables, 78t
  project implementation
    partner organizations, 73—74
    Stantec Consulting Services Inc.
      (Stantec), 74—75
    study locations, 75f
    task orders, 77
    United States Geological Survey
      (USGS) Arctic report, 73—74
  project planning
    IDIQ government contract structure,
      74f
    multiple solicitation approach, 72—73
    National Oceanographic Partnership
      Program (NOPP), 71—72
    objectives, 72
    request for proposals (RFP), 73
    solicitations and awards, 72, 72f
    task orders, 73
  request for proposals, 83—84
  solicitations, 83—84
  sustainability, 77—79
  trust and continuity, 82—83
Marine citizen science
  active participation programs, 88
  Coastal Observation and Seabird Survey
    Team (COASST). *See* Coastal
    Observation and Seabird Survey
    Team (COASST)
  data-driven projects, 88
  digital imagery, 88—89
  partnerships, 104
  Reef Education Environmental
    Foundation (REEF), 89
  sample collection, 89—90
  volunteer-based marine sensor
    deployment programs, 88
Marine research partnerships
  adaptive process, 168—173
  analysis, 168—176
  communication, 175
  ethical outcomes, 180
  flexibility, 174—175
  leadership, 173—174
  operational goals, 179—180
  programmatic elements, 168
  resilience and sustainability,
    176—178, 179f

scientific research products, 167–168
social process, 167–168
time needs, 175–176
trust-building, 174
vision, 173
Minerals Management Service (MMS), 112, 132
Multi-Agency Rocky Intertidal Network (MARINe)
   beginning and structure, 114–116
   Bureau of Land Management (BLM) program, 112
   communication, 120–121
   communication structure, 115f
   consensus-based governance process, 116–117
   continuity, 121–123
   core protocol data, 117–118
   culture of care, 110–111, 111b
   data consistency, 118–119
   first mission statement, 114–115
   foundations, 112–117
   funding, 123–124
   membership requirements, 116
   mission and goal, 109
   National Park Service (NPS), 112
   partnership, 125–126
   protocol development, 117–119
   Science Panel, 115–116
   shared database, 119–120
   Shoreline Inventory Project, 112–113
   survey sites, 110f
   sustainable practices, 117–124
   timeline of events, 113b

N

Nansen Legacy project
   Arctic Ocean and blue water masses, 36f
   Barents Sea region, 33–34
   bottom-up approach, 46–47
   challenges and activities
      architecture, 39
      Board meetings, 39–40
      competitive principle, 40
      Norwegian partners, 40
   cooperation, 44
   division of labor, 44
   funding, 42
   future Arctic Ocean, 47–49
   goals, 37–38
   national research teams, 45
   Norway's Arctic fisheries and environmental strategy, 35–37
   Norwegian CABANERA project, 34–35
   ProMare research project, 34–35
   recruitment, sustainable Arctic future, 43
   research plan
      industry and international partners, 41
      quality control, 41–42
National Biological Survey (NBS), 135–136
National Defense Authorization Act, 184
Nationally Determined Contributions (NDCs), 201–202
National Oceanographic Partnership Program (NOPP), 59, 71–72, 136–137, 184
National Park Service (NPS), 112
National Science Foundation (NSF), 1
National Seabird Program, 96
National Studies List (NSL), 133
North Pacific Research Board (NPRB), 1
North Slope Science Initiative (NSSI), 188

O

Olympic Coast National Marine Sanctuary (OCNMS), 94–95

P

Partnership for Regional Ocean Governance (PROG), 202
Policy sources and impacts, 189–191
ProMare research project, 34–35
Protocol drift, 116–117

Q

Quality assurance/quality control (QAQC) measures, 90

R

Reef Education Environmental Foundation (REEF), 89
Regional Ocean Management Organizations (ROMOs), 202
Repetitive flooding
　adaptation possibilities, 156
　in Alaska, 156–157
　coastlines, 155
　co-production
　　academic and nonacademic products, 163
　　approval and consent, 159–160
　　Arctic Social Science Program (ASSP), 158–159
　　budget allocation, 160
　　data collection mechanisms, 160–161
　　data collection strategy, 162
　　"decolonized" interview session, 161
　　informal relationship, 158
　　Institutional Review Board (IRB), 163
　　local collaborators, 158
　　pandemic experiences, 161
　group-relocation schemes, 155–156
　Indigenous communities, 155–156
　structural adjustment, 156
Research policies and partnerships
　academic research partnerships, 184–185
　case studies, 187–189
　foundational principles, 193–194
　global partnership, 191
　implementation stage, 184
　institutional goals, 193
　knowledge production process, 185
　knowledge systems, 184–185
　policy document, 184
　power disparities and dynamics, 185–186
　scenario planning methods, 194
　science and government institutions, 186–187
　scientific questions, 185
　societal support, 191
　sources and impacts, 189–191, 189f
　　endogenous source, 189–190
　　exogenous sources, 189–190
　procurement vehicles, 190
　US Federal Acquisition Regulations (FAR), 190
　sponsors and funders, 194

S

Scoping process, 18–19
Shoreline Inventory Project, 112–113
Social process, 167–168
Study Development Plan (SDP), 133
Sustainability
　argo program, 64–66
　bering Sea project, 10–11
　Coastal Observation and Seabird Survey Team (COASST), 102–104
　coastal Observation and Seabird Survey Team (COASST), 102–104
　Marine Arctic Ecosystem Study (MARES), 77–79
Sustainability Research and Innovation (SRI), 26

T

Thematic Program Office (TPO), 19, 21
Tropical Ocean Global Atmosphere Project (TOGA), 56–57
Trust-building, 174
Turtles Uniting Researchers and Tourists (TURT), 88–89

U

UN Convention for the Law of the Sea (UNCLOS), 201
University of California, Santa Barbara (UCSB), 112–113
US Federal Acquisition Regulations (FAR), 190
US Geological Survey (USGS), 134

W

Washington Department of Fish and Wildlife (WDFW), 96
World Meteorological Organization (WMO), 62

Z

Zooniverse, 88–89

Printed in the United States
by Baker & Taylor Publisher Services